STARS

**A GUIDE TO THE CONSTELLATIONS,
SUN, MOON, PLANETS,
AND OTHER FEATURES OF THE HEAVENS**

HERBERT S. ZIM, Ph.D., Sc.D.

and

ROBERT H. BAKER, Ph.D., D.Sc.

Illustrated by

JAMES GORDON IRVING

 GOLDEN PRESS • NEW YORK
Western Publishing Company, Inc.
Racine, Wisconsin

FOREWORD

A series of books on nature should include one on the stars and planets. All other aspects of nature—birds, flowers, rocks, and trees—are details in the great system that encompasses the heavens. Nothing else in nature can arouse the feelings of wonder that are provoked by an eclipse, a meteor shower, or even a close look at our nearest neighbor, the moon.

The artist, James Gordon Irving, worked with skill and imagination. His wife, Grace Crowe Irving, assisted in research. David H. Heeschen of the Harvard Observatory and Ivan King of the University of Illinois Observatory helped with data and tables. Paul Lehr, of the National Oceanic and Atmospheric Administration, checked text involving meteorology. Hugh Rice of the Hayden Planetarium gave helpful advice, and our seasonal constellation charts owe much to his projections. Dorothy Bennett, for many years a member of the Hayden Planetarium's staff, contributed greatly to our editorial planning. Isaac Asimov, Joe and Simone Gosner are to be credited for the latest revisions.

Acknowledgment is due the Lowell, Hale (Mt. Wilson and Mt. Palomar), Lick, and Yerkes observatories and to NASA for the use of photographs.

R. H. B.
H. S. Z.

GOLDEN®, A GOLDEN GUIDE®, GOLDEN PRESS® and GOLDENCRAFT® are trademarks of Western Publishing Company, Inc.

© Copyright 1975, 1956, 1951 by Western Publishing Company, Inc. All rights reserved, including rights of reproduction and use in any form or by any means, including the making of copies by any photo process, or by any electronic or mechanical device, printed or written or oral, or recording for sound or visual reproduction or for use in any knowledge retrieval system or device, unless permission in writing is obtained from the copyright proprietor. Produced in the U.S.A. by Western Publishing Company, Inc. Published by Golden Press, New York, N.Y. Library of Congress Catalog Card Number: 75-314330.
ISBN 0-307-63507-4

CONTENTS

This is a book for the novice, the amateur, or anyone who wants to enjoy the wonders of the heavens. It is a field guide, with information to help you understand more fully what you see. Use this book when you are watching the stars, constellations, and planets. Thumb through it at odd moments to become familiar with sights you may see; carry it along on trips or vacations.

OBSERVING THE SKY	4
Activities for the Amateur	7
The Universe and the Solar System	12
The Sun and Sunlight	16
Telescopes	28
STARS	31
Classification	36
Star Types	38
Galaxies	42
CONSTELLATIONS	50
North Circumpolar Constellations	52
Constellations of Spring	62
Constellations of Summer	70
Constellations of Autumn	80
Constellations of Winter	88
South Circumpolar Constellations	98
THE SOLAR SYSTEM	102
The Planets	104
Locating the Visible Planets	124
Comets	126
Meteors	129
How to Observe Meteors	132
The Moon	136
Eclipses	150
APPENDIX (tinted pages)	156
List of Constellations	156
Objects for Observation	158
INDEX	159

Egyptian Pyramids

OBSERVING THE SKY

Stars and planets have attracted man's attention since earliest times. Ancient tablets and carvings show that movements of planets were understood before 3000 B.C. Legend says two Chinese astronomers who failed to predict an eclipse correctly in 2136 B.C. were put to death. The Egyptians placed their pyramids with reference to the stars. The circles of stone at Stonehenge may have been used to keep track of lunar eclipses. Astronomy is indeed the oldest science, yet its importance increases as scientists turn to the stars to study problems of physics which they cannot tackle directly in the laboratory.

As far back as history records, there were professional astronomers—long before there were professional zoologists and botanists. The Egyptians, Chinese, and Europeans had court astronomers. Their work often involved trying to predict future events, but their system, though considered unscientific today, involved observation and recording of facts about stars and planets. These early astronomers, as well as those of today, made remarkable discoveries that changed man's outlook on the world and himself.

There has always been, too, an army of amateurs studying and enjoying the stars. Some make practical use of their knowledge—sailors, pilots, surveyors, but most study the heavens out of sheer interest and curiosity.

WHY LOOK? • The stars can tell you time, direction, and position. These are about their only practical use to an amateur. More important is the satisfaction one finds in recognizing the brightest stars and planets. To see and to recognize Leo in the eastern sky is akin to seeing the first robin. And, as you learn more about the stars and the variety of other celestial objects, the more the wonder of the heavens grows.

WHERE TO LOOK • Star-gazing has no geographic limits. Some stars can even be seen from brightly lit, smoky city streets, but the less interference from lights or haze the better. An ideal location is an open field, hill, or housetop where the horizon is not obscured by trees or buildings. However, buildings or a hill may also be used to screen off interfering lights, and although you may see less of the sky this way, you will be able to see that part of it better.

WHEN TO LOOK • Only the brighter stars and planets are visible in full moonlight or soon after sunset. At these times the beginner can spot them and learn the major constellations, without being confused by myriads of fainter stars. On darker nights, without moonlight, one may observe minor constellations, fainter stars, nebulae, and planets. Stars and planets visible at any given hour depend on time of night and season of the year. As the earth rotates, new stars come into view in the eastern sky as the evening progresses. Late at night one can see stars not visible in the evening sky.

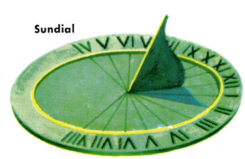

Sundial

until several months later. The seasonal star charts (pp. 64-65, 72-73, 82-83, 90-91) and planet tables (pp. 124-125) show the location of major celestial objects at various times of the year. See check list, p. 158.

HOW TO LOOK • First, be comfortable. Looking at stars high above the horizon may cause a stiff neck and an aching back; so use a reclining chair, a couch, or a blanket spread on the ground. Remember—ground and air may be unexpectedly cold at night; warm clothing, even in summer, may be needed. How to look also involves a method of looking. The section on constellations (pages 50-101) gives suggestions. After you have become familiar with the more common stars, constellations, and planets, a systematic study may be in order—perhaps with field glasses. By that time your interest may lead you to some of the activities suggested on the following pages.

EQUIPMENT • You need no equipment, except your eyes, to see thousands of stars. This book will point the way to hours of interesting observation with your eyes alone. Later you will find your enjoyment greatly enhanced by the use of field glasses (6- to 8-power) such as those used in bird study. With these you can see vastly more—details on the moon, moons of Jupiter, many thousands of stars, star clusters, double stars, and nebulae.

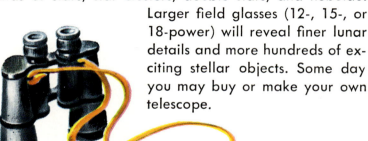

Larger field glasses (12-, 15-, or 18-power) will reveal finer lunar details and more hundreds of exciting stellar objects. Some day you may buy or make your own telescope.

ACTIVITIES FOR THE AMATEUR

ENJOYING THE STARS • It is worth repeating that night-by-night observing, studying, and enjoying the stars is the activity that can mean the most to most people. No equipment and little preparation are needed. This book (see p. 158) and sources of information suggested (p. 11) will help.

Zeiss, Projection Planetarium

IDENTIFICATION • The enjoyment of stars involves some practice in identification. Knowing two dozen constellations and a dozen of the brightest stars is often enough. A systematic study of stars, the identification of lesser constellations, and the location and study of clusters and nebulae demand more intensive efforts. A serious amateur will benefit by knowing nearly all the constellations and bright stars before going deeper into any phase of astronomy.

FOLLOWING THE PLANETS • The planets, moving along in their orbits, are constantly changing their positions. Even the beginner can become familiar with the movements of planets—can recognize them, and predict which way they will travel. Knowing the planets is as important and as enjoyable as knowing the stars.

MUSEUMS • Many museums have astronomical exhibits worth seeing. These may include meteorites, photographs of stars and planets, and sometimes working models of

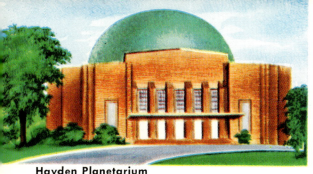
Hayden Planetarium

the solar system. Museums may be found at universities, observatories, planetariums, or governmental institutions. Inquire locally or when traveling concerning museums in the area that may offer astronomical exhibits.

OBSERVATORIES • These are the sites of the great telescopes and the places where professional astronomers work. When work is going on, astronomers cannot be disturbed. But most observatories are open to the public at specified hours. Some offer a schedule of public lectures.

PLANETARIUMS • In New York, Chicago, Pittsburgh, Philadelphia, Chapel Hill (N.C.), Boston, Los Angeles and San Francisco are located the large projectors that show images of stars and planets on a darkened dome. Hundreds of people can watch these artificial stars while a lecturer points them out and describes other features of the heavens. Many other planetariums are equipped with smaller but very effective projectors.

CLUBS AND ASSOCIATIONS • Amateur astronomers often band together to share their experience and their interests. Clubs are found in most large cities and in many smaller ones. At meetings, a lecture or discussion is usually followed by a period of observing through telescopes. Some clubs work on co-operative projects in which the members share in some

Copernican Planetarium

scientific investigation. Visitors are usually welcome, and membership is commonly open to anyone who is interested.

Any person interested in the stars will enjoy the activities described above. The thrill of knowing the planets or visiting a planetarium can be shared by young and old alike. Through such activities you may become a serious amateur. The serious amateur may be anyone from a youth in high school to a business man. Such amateurs spend much of their time working on an astronomical hobby. They often become experts; some have made important discoveries. Professional astronomers are glad to have the help of trained amateurs, and several fields of astronomical research are manned largely by them. Amateur activities that demand greater skill and experience offer greater rewards in the satisfaction they provide to serious amateurs. Here are some:

Grinding a Mirror

TELESCOPE MAKING • Making a telescope requires time and patience. But in the end you have an instrument costing only a small fraction of its worth, plus the fun of having made it. The telescopes made by amateurs are usually of the reflecting type, with a concave mirror instead of a lens for gathering light. Telescope-making kits, including a roughly finished glass "blank" for the mirror, other telescope parts, and complete instructions, are available from some optical-supply firms. The final testing and finishing of the mirror require special skill and often the help of someone with experience.

Armillary Sphere Once Used to Demonstrate Celestial Motions

OBSERVING METEORS
• Meteors or shooting stars (pp. 129-133) often occur in well-defined showers. Careful observation and plotting of the paths of meteors yield information of scientific value. A number of groups of amateurs are engaged in observing meteors, and any interested amateur or group of amateurs can join. Contact the American Meteor Society, 521 N. Wynnewood Ave., Narberth, Pennsylvania 19072.

OBSERVING VARIABLE STARS • Amateurs with telescopes have done unusual work in this advanced field. Studies of these stars are co-ordinated by the American Association of Variable Star Observers, 4 Brattle St., Cambridge, Massachusetts 02100. The director of the Association will be glad to furnish qualified amateurs with details about this work.

STELLAR PHOTOGRAPHY • Photographing the stars and other heavenly bodies is not difficult. Excellent pictures have been taken with box cameras set firmly on a table. But pictures of faint objects must be taken with a telescope or with a special camera adjusted to compensate for the earth's motion. Photography is an important tool of astronomers—one which the amateur can use to good advantage.

MORE INFORMATION • This book is a primer to the sky and can only introduce a story which is more fully told in many texts and popular books on astronomy.

BOOKS:

Baker, Robert H., *Astronomy*, D. Van Nostrand Co., New York, 1964. This college text is for the serious student who wants detailed facts and modern theories.

Baker, Robert H., *When the Stars Come Out*, 1954, and *Introducing the Constellations*, 1957. Viking Press, New York. This pair of books serves to introduce the stars, meeting the needs of persons without scientific background. They are well illustrated.

Bernhard, Bennett, and Rice, *New Handbook of the Heavens*, Whittlesey House, New York, 1954. A fine book, bridging the gap between popular volumes and texts. It stresses things to do and see. Also in a pocket edition by Signet Books.

Mayall, Mayall and Wyckoff, *The Sky Observer's Guide*, A Golden Handbook, Golden Press, New York, 1965. An introductory book for the layman with detailed maps of the heavens.

Olcott, Mayall, and Mayall, *Field Book of the Skies*, G. P. Putnam's Sons, New York, 1954. The revised edition of a popular and practical guide for the amateur observer.

PERIODICALS:

Sky and Telescope. This is the outstanding magazine for the amateur. Sky Publishing Corp., 49-50-51 Bay State Road, Cambridge, Mass. 02138.

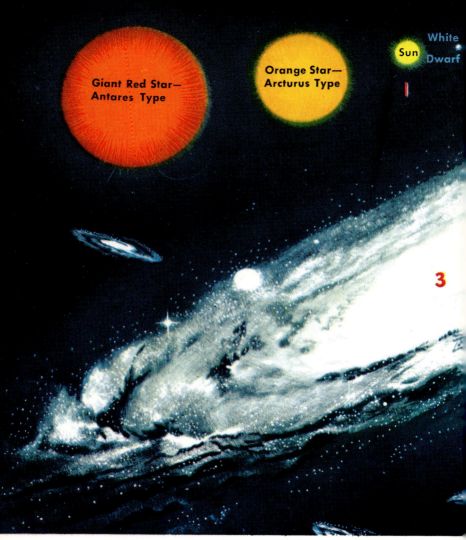

OUR UNIVERSE is so vast that its limits are unknown. Through it are scattered millions of galaxies of various sizes and shapes. In a galaxy like one shown here (3), our sun and the earth are located (see p. 42). Galaxies contain hundreds of millions, even hundreds of billions, of stars of many types (1), ranging from red supergiants

less dense than the earth's atmosphere to white dwarfs hundreds of times denser than lead. Stars on the average are spaced several light years apart; but there are some clusters (2), more closely packed toward the center that contain perhaps half a million stars in all. Planets may revolve around many of the stars.

OUR SOLAR SYSTEM is located halfway from the center of our galaxy — the Milky Way. Around the sun revolve the nine planets with their 32 satellites; also hundreds of asteroids and swarms of meteors. Here we see the planets (1) in their orbits around the sun (see pp.

102-105) and (2) in the order of their size. The asteroid Ceres is compared (3) to Texas for size, and the moon is compared (4) to the United States. A comet's orbit (5) appears in red. Our solar system may be only one of billions in the universe.

Solar Prominences compared to Size of Earth

THE SUN is the nearest star. Compared to other stars it is of just average size; yet if it were hollow, over a million earths would easily fit inside. The sun's diameter is 860,000 miles. It rotates on its axis about once a month. The sun is gaseous; parts of the surface move at different speeds. The sun's density is a little under 1½ times that of water.

The sun is a mass of incandescent gas: a gigantic nuclear furnace where hydrogen is built into helium at a temperature of millions of degrees. Four million tons of the sun's matter is changed into energy every second. This process has been going on for billions of years, and will continue for billions more.

The sun's dazzling surface, the photosphere, is speckled with bright patches and with dark sunspots (pp. 22-23). Rising through and beyond the chromosphere, great prominences or streamers of glowing gases shoot out or rain down. The corona, which is the outermost envelope of gases, forms a filmy halo around the sun.

It is unsafe to observe the sun directly with the naked eye or binoculars. Use a special filter, a dark glass, or a film negative to protect your eyes. When a telescope is used, project the sun's image on a sheet of paper.

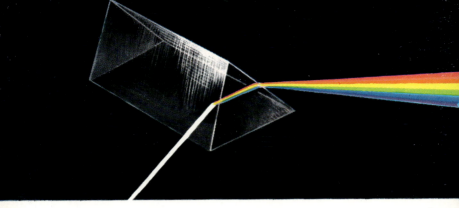

How a Prism Breaks Up a Beam of Sunlight into Its Component Colors

SUNLIGHT • Every square yard of the sun's surface is constantly sending out energy equal to the power of 700 automobiles. About one two-billionth of this, in the form of sunlight, reaches us. Sunlight is a mixture of colors. When it passes through a glass prism, some of the light is bent or refracted more than other portions. Light leaving the prism spreads out into a continuous band of colors called a spectrum. Colors grade from red, which is bent least, through orange, yellow, green, and blue to violet, which is bent most.

The spectrum is crossed by thousands of sharp dark lines. These indicate that some light was absorbed as it

Fraunhofer lines
Invisible Ultraviolet   G   F   E

passed through the cooler gases above the sun's surface. These gases absorb that part of the sunlight which they would produce if they were glowing at a high enough temperature. Thus a study of the dark lines in the solar spectrum (called Fraunhofer lines, after their discoverer) gives a clue to the materials of which the sun is made. Of the 92 "natural" elements on the earth, 61 have been found on the sun. The rest are probably present also. From the shifting of spectral lines, astronomers can measure the rotation of the sun and the motions of stars. They can detect magnetic fields from spectral lines and can determine a star's temperature and its physical state.

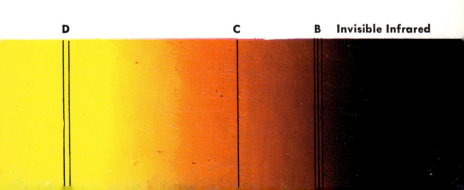

RAINBOWS are solar spectra formed as sunlight passes through drops of water. Rainbows may be seen when a hose is adjusted to a fine spray. The drops act like prisms, refracting sunlight to produce the spectrum.

A single, or primary, rainbow has red on the outside, violet inside. The center of the arc, 40 degrees in radius, is always on a line with the observer and the sun. When you see a bow, the sun is behind you. Sometimes a secondary rainbow forms outside the primary. It is fainter, with colors reversed—red inside, violet outside. The secondary bow

A Rainbow Is a Spectrum

forms from light reflected twice within drops. Light may be reflected more than twice, so occasionally up to five bows are seen.

Another type of bow — red, or red and green — may appear with primary and secondary bows.

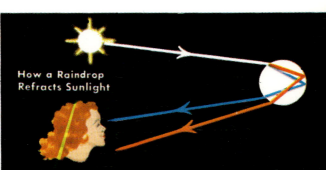

How a Raindrop Refracts Sunlight

SUNSPOTS often appear on the sun's photosphere—appearing as dark, sculptured "holes" in contrast to the bright white surface. These sunspots are sometimes so large they can be seen with the unaided eye (through a dark glass for protection, of course), and are most easily observed when the sun is low on the horizon. The use of field glasses or a small telescope helps, but the safest method of observation is to study photographs. The dark center, or umbra, of a sunspot varies from a few hundred to over 50,000 miles across. This is surrounded by a less dark area, a penumbra, that often doubles the size of the sunspot. As the sun rotates, new sunspots come into view. Most persist for a week or so, but the maximum duration is from three to four months.

The number of sunspots varies in cycles of about 11 years; first increasing steadily until hundreds of groups are seen annually, then gradually decreasing to a minimum of about 50 groups. At the beginning of a cycle the sunspots appear about 30° north and south of the sun's equator. As the cycle progresses, they develop closer to the equator and the zone of activity extends from 10 to 20 degrees on either side of it. The 11-year

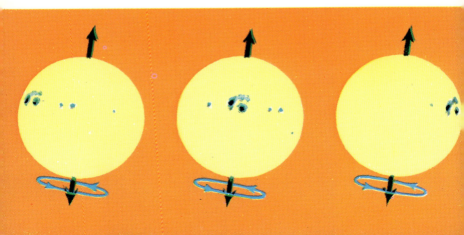

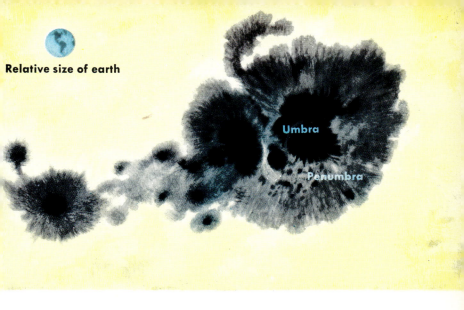

Relative size of earth

Umbra

Penumbra

cycle is really an average value. Mysteriously waxing and waning, the exact length of the cycle can be as short as nine years or as long as sixteen. Some astronomers have pointed out that the 11-year cycle seems to be part of a larger 22-year cycle in which the entire magnetic field of the sun may reverse itself.

Sunspots seem to be giant magnetic storms on the sun's surface, which may be caused by deeper, periodic changes. They occur in groups which grow rapidly and then slowly decline. The gases in the sunspot (about 8000°F) are cooler than the rest of the sun's surface (about 11,000°F); hence they appear darker. Actually, if a large sunspot could be isolated in another part of the sky, it would appear as bright as a hundred full moons. Sunspots have strong magnetic fields. Radiation from "solar flares" near them interacts with the upper levels of the earth's atmosphere and interrupts short-wave radio transmission; it is also likely to cause an increase in auroras (pp. 24-25.)

AURORAS OR NORTHERN LIGHTS • The shifting, glowing, diffuse light of an aurora is hard to describe. Yellow, pink, and green lights come and go; arcs of light start at the horizon and spread upward; streamers and rays extend toward the zenith. Auroras last for hours, and often all through the night. They are seen in the north and middle northern latitudes and in the arctic. A similar display is seen in the southern latitudes. Auroras occur from about 60 to 600 miles up in the air. At these heights, so little air remains that space is almost empty like a

vacuum, or the inside of a neon light. The shifting glow of the aurora is essentially electrical and somewhat similar to the light from the neon signs along Main Street.

The sun's radiation on the rare gases of the upper atmosphere is what causes auroras. Auroras may be due to charged particles that come from solar flares near sunspots. The fact that auroras seem to center around the earth's magnetic poles emphasizes their electrical character. A few days after a large new sunspot group develops, an auroral display is likely to occur.

THE SKY FROM SUNRISE TO SUNSET • As the sun's rays pass through the earth's atmosphere, some are scattered, and a play of colors results. Blue rays are scattered most, and therefore a clear sky is typically blue. Yellow rays are scattered less than blue; thus the sun itself, so long as it is well above the horizen, looks yellow. But just after sunrise and just before sunset the sun is reddish. At these times the sharply slanting sun's rays must travel a longer path through the atmosphere, and more of the blue and yellow rays are scattered. The red rays, which are scattered least, come through in the largest numbers,

Sunlight Passes Through a Thicker Layer of Air at Sunrise and Sunset

giving the sun its reddish hue. If there are clouds and dust in the air, many of the red rays which filter down into the lower atmosphere are reflected, and large areas of the sky may be reddened.

Because of the bending or refraction of light, which is greater when the sun is near the horizon, you can actually see the sun for a few minutes before it rises and after it sets. Daylight is a bit longer for this reason. The closer to the horizon, the greater the refraction at sunrise or sunset. Hence, as refraction elevates the sun's disc, the lower edge is raised more than the upper. This distorts the sun, just as it is rising or setting, giving it an oval or melon-shaped appearance.

Twilight is sunlight diffused by the air onto a region of the earth's surface where the sun has already set or has not risen. It is generally defined as the period between sunset and the time when the sun has sunk 18 degrees below the horizon—that is, a little over an hour.

THE TELESCOPE was first put to practical use by Galileo in 1609. Since then, it has extended man's horizons farther and has challenged his thinking more than any other scientific device. The telescope used by Galileo, the best-known kind, is the refracting telescope, consisting of a series of lenses in a tube. In a simple refractor, two lenses are used, but commonly others are added to correct for the bending of light that produces a colored halo around the image. The largest refracting telescopes are one with a 40-inch lens at the Yerkes Observatory in Wisconsin, and a 36-inch one at Lick Observatory in California.

The simple reflecting telescope has a curved mirror at the bottom of the tube. This reflects the light in converging rays to a prism or diagonally placed mirror, which sends the light to the eyepiece or to a camera mounted at the side of the tube. Since mirrors can be made larger than lenses, the largest astronomical telescopes are reflectors (see page 30). Reflectors with mirrors up to 8

40-Inch Refractor, Yerkes Observatory

Refracting Principle

0-Inch Reflector, Palomar Mountain, Calif.

Reflecting Principle

inches in diameter are made by amateurs as the best simple, low-cost telescope. A special type of reflector, the Schmidt, permits rapid photographing of the sky and has become one of the major astronomical tools. Also of growing importance are the radio telescopes — giant, saucer-shaped instruments which receive radio waves from outer space. Clouds of optically invisible neutral hydrogen, abundant in the spiral arms of our galaxy, can be traced by their radio emissions. Radio astronomy is a young science.

THE LARGEST TELESCOPE of the reflector type is on Palomar Mountain, near San Diego, Calif. Its 200-inch (16.6-foot) mirror is a marvel of scientific and engineering skill. The great disc of pyrex glass was cast with supporting ribs to bear its weight. It is 27 inches thick and weighs 14½ tons. Yet because of its design, every part is within two inches of the air—permitting the mirror to expand and contract uniformly with changes in temperature. The great piece of glass has been polished to within a few millionths of an inch of its calculated curve. Despite its great weight it can be tilted and turned precisely without sagging as much as the thickness of a hair. The mirror gathers about 640,000 times as much light as the human eye. With it, astronomers photograph stars six million times fainter than the faintest stars you can see, and galaxies over two billion light years away.

STARS

Stars are suns: heavenly bodies shining by their own light and generally so far away from us that, though moving rapidly, they seem fixed in their positions.

NUMBERS OF STARS • On the clearest night you are not likely to see more than 2,000 stars. With changing seasons, new stars appear, bringing the total visible during the year to about 6,000. A telescope reveals multitudes more. The total in our galaxy runs into billions, but even so, space is almost empty. Were the sun the size of the dot over an "i," the nearest star would be a dot 10 miles away, and other stars would be microscopic to dime-size dots hundreds and thousands of miles distant.

DISTANCES OF STARS • The nearest star, our sun, is a mere 93 million miles away. The next nearest star is 26 million million miles—nearly 300,000 times farther than the sun. For these great distances, miles are not a good measure. Instead, the light year is often used. This is the distance that light travels in one year, moving at 186,000 miles per second: nearly 6 million million miles. On this scale the nearest star (excluding the sun) is 4.3 light years away. Sirius, the brightest star, is 8.8 light years off. Other stars are hundreds, thousands, and even millions of light years away.

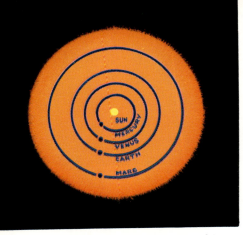

Antares Is Larger Than Mars' Orbit

STARLIGHT • All stars shine by their own light. This light may be produced by nuclear reactions similar to those of the hydrogen bomb. When the element hydrogen is transformed into helium, which happens on most stars, about 1 per cent of its mass (weight) is changed into energy. This energy keeps the temperature in the star's interior at millions of degrees. At the surface the temperature varies from about 5,500 degrees F. to over 55,000 degrees, depending on the kind of star. One pound of hydrogen changing to helium liberates energy equal to about 10,000 tons of coal. In a single star the energy released in this way requires the transformation of millions of tons of matter per second.

STAR BRIGHTNESS • The sun is about average in size and brightness. Some stars are up to 600,000 times as bright as the sun; others are only 1/550,000; most are between 10,000 and 1/10,000 times as bright as the sun. The brightness of a star you see depends on its distance and on its real or absolute brightness. See pp. 34-35.

STAR SIZE • Most stars are so distant that their size can only be measured indirectly. Certain giant red stars are the largest. Antares has a diameter 390 times that of the sun, others even larger. Among the small stars are white dwarfs, no larger than planets. The smallest are neutron stars that may be no more than ten miles across.

DENSITY OF STARS • The densities or relative weights of stars vary considerably. Actually all stars are masses of gas—but gas under very different conditions from those we usually see. Giant stars such as Antares have a density as low as 1/2000 of the density of air. More usual stars have a density fairly close to that of the sun. White dwarfs are so dense that a pint of their material would weigh 15 tons or more on earth. The companion to Sirius is 25,000 times more dense than the sun. Neutron stars are billions of times denser.

MOTIONS OF STARS • Our sun is moving about 12 miles per second toward the constellation Hercules. Other stars are moving too, at speeds up to 30 miles per second or faster. Arcturus travels at 84 miles per second. Many stars are moving as parts of systems or clusters. One such system, including stars in Taurus, is moving away at about 30 miles per second. Some stars consist of two or more components (see p. 38) which revolve around a common center as they move together through space. The stars in a constellation do not necessarily belong together; they may be of widely differing distances from the earth and may be moving in different directions at different speeds.

COLOR OF STARS varies from brilliant blue-white to dull reddish, indicating star temperature (pp. 36-37)—a factor in star classification. Close observation is needed to see the range of colors in the night sky.

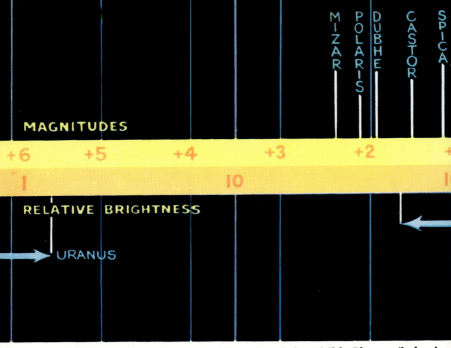

Brightness of Some Major Stars (above) and the Visible Planets (below)

STAR MAGNITUDES • Brightness of stars is measured in terms of "magnitude." A 2nd-magnitude star is 2.5 times as bright as a 3rd, and so on throughout the scale, so that a 1st-magnitude star is 100 times as bright as a 6th. Stars brighter than 1st magnitude have zero or minus magnitude. On this scale the magnitude of the planet Venus is —4; it is 10,000 times as bright as a 6th-magnitude star, which is the faintest that the unaided eye can see. The sun's magnitude is —27.

The brightness of a star as we see it depends on two factors: its actual, or absolute, brightness and its distance from us. If one factor is known, the other can be computed. This relationship makes it possible to measure the distance of remote galaxies (p. 39).

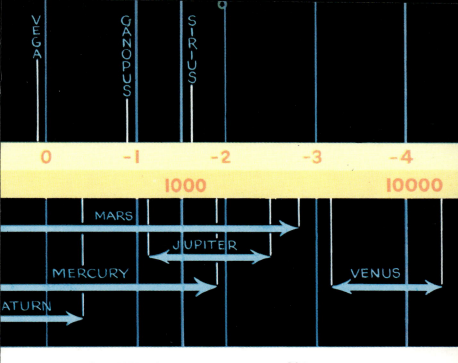

THE BRIGHTEST STARS

Name	Constellation	Magnitude as seen	Distance (light years)	Absolute magnitude
Sirius	Canis Major	—1.4 d	9	+1.5
*Canopus	Carina	—0.8	540	—5.0
*Alpha Centauri	Centaur	—0.3 d	4	+4.1
Arcturus	Boötes	0.0	32	+0.2
Vega	Lyra	0.1	26	+0.6
Capella	Auriga	0.1	45	—0.5
Rigel	Orion	0.1 d	650	—6.5
Procyon	Canis Minor	0.4 d	11	+2.7
*Achernar	River Eridanus	0.6	140	—2.6
*Beta Centauri	Centaur	0.7	140	—2.5
Betelgeuse	Orion	0.7	270	—4.1
Altair	Aquila	0.8	16	+2.4
*Alpha Crucis	Southern Cross	0.8 d	160	—2.7
Aldebaran	Taurus	0.9 d	68	—0.6
Spica	Virgo	1.0	230	—3.3
Antares	Scorpius	1.0 d	410	—4.0
Pollux	Gemini	1.1	34	+1.0
Fomalhaut	Southern Fish	1.2	23	+2.0
Deneb	Cygnus	1.2	1500	—7.0
Regulus	Leo	1.3 d	86	+1.0
*Beta Crucis	Southern Cross	1.3	470	—4.5

*Not visible at 40° N. latitude. d Double stars; combined magnitude given.

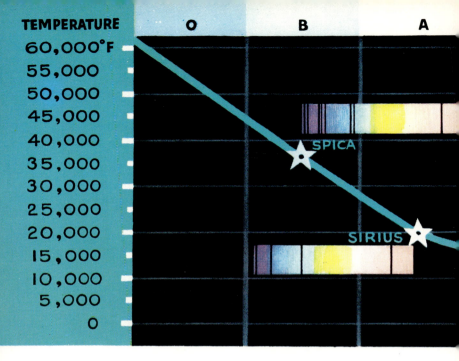

CLASSIFICATION OF STARS • Most of the stars you see can be classified into seven groups according to the stars' spectra. These, in turn, depend mainly on the temperature of the stellar atmosphere. The spectra of low-temperature stars show that some simple chemical compounds are present. As the temperatures of stars increase, the spectra reveal that fewer molecules occur, and that the atoms making up the elements that are present become excited and ionized. Ionized atoms are those which have lost one or more electrons. On the basis of studies of thousands of spectra, stars are arranged in seven classes: O, B, A, F, G, K, and M. For more detailed study, astronomers divide each class into ten sub-classes, as A0, B3, or G5. Over 99 per cent of the stars fit into this classification. Four other classes (W, R, N, S) are used for stars not fitting the seven main groups.

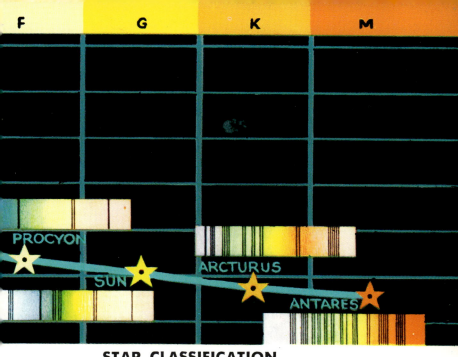

STAR CLASSIFICATION

Star class	Approx. Temp. (degrees F.)	Color	Spectral character	Examples
O	over 55,000	Blue-white	Gases strongly ionized	Iota Orionis (in sword)
B	36,000	Blue-white	Strong neutral helium	Rigel Spica
A	20,000	White	Hydrogen predominant	Sirius Vega
F	13,500	Yellowish white	Hydrogen decreasing; metals increasing	Canopus Procyon
G	11,000	Yellow	Metals prominent	Sun Capella
K	7,500	Orange	Metals surpass hydrogen	Arcturus Aldebaran
M	5,500	Red	Titanium oxide present—violet light weak	Betelgeuse Antares

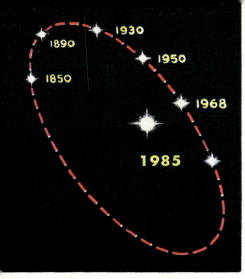

Pair of Double Stars: Castor

STAR TYPES

DOUBLE STARS • Over a third of all known stars are double, or "binary." The components of a few can be seen with the unaided eye; thousands can be "separated" with a telescope; thousands more are detected by the spectroscope. Some stars have three or more components; Castor has six—three doubles. The main, mutually revolving pair was closest together in 1968—about 55 times the distance of the earth from the sun.

Mizar, at the curve of the Big Dipper's handle, has a faint companion. Mizar itself is a telescopic double and the brightest component is a spectroscopic double star. When the two stars are in line, the spectra coincide. When, as they revolve, one approaches us as the other moves away, the spectrum lines are doubled. Capella and Spica are also spectral double stars.

As double stars revolve, one may eclipse the other, causing reduced brightness. Best-known of eclipsing double stars is Algol, in Perseus. It waxes and wanes at intervals of about three days. The eclipsing stars are 13 million miles apart. Their combined magnitude varies from 2.3 to 3.4.

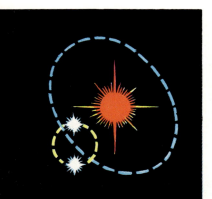

Triple Star System

VARIABLE STARS are those that fluctuate in brightness. Most dramatic are the exploding stars, or novae. These dense, white stars rapidly grow in brilliance, up to 100,000 times or more, then fade away. Other variables change less drastically regularly or irregularly. Some red giants and supergiants

Nova of 1572 in Cassiopeia

vary from 4 to 10 magnitudes over a few months to two years. Mira, in Cetus the Whale, is a famous long-period variable that shows extreme changes of brightness.

The variables known as classical Cepheids vary in brightness over periods of one day to several weeks. The distance of any of these Cepheids can be readily estimated because of the definite relationship between its variation period and its absolute magnitude. By measuring the period, the absolute magnitude can be determined, and by comparing the absolute magnitude with the apparent magnitude, the distance can be estimated. These Cepheids are important in the calculation of distances of star clusters and galaxies (pp. 40-43). In recent years, early information about classical Cepheids has indicated that estimates of the distances of galaxies beyond our own should be at least tripled.

Eclipsing Binary, Algol, with Magnitude Changes

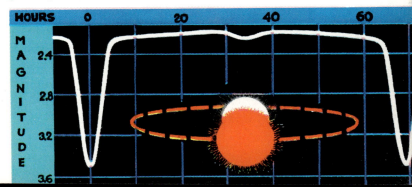

The Globular Cluster M13 in Hercules — Palomar

STAR CLUSTERS are groups of stars relatively close to one another and moving together as a stellar system. Clusters are of two types—open and globular. Some 300 open clusters occur in our galaxy (p. 42). Some are fine objects to observe with binoculars (p. 158). One close "moving cluster" includes most of the stars in the Big Dipper. Another open cluster includes approximately 150 stars in and around Taurus, the Bull. Some open clusters are easier to recognize, as the Praesepe cluster in Cancer, the Coma Berenices cluster, and the double cluster of Perseus. Most open clusters are in or near the Milky Way.

Open Cluster: Pleiades with Nebulosity　　　　　　　　　Yerkes

    Globular clusters are much more compact, often more distant. The brightest appear as dim, hazy spots; few can be seen with the unaided eye. About 100 have been found in our galaxy, and many more in others. The great Hercules cluster M13 (it was No. 13 in the astronomical catalogue of Messier) is striking in large telescope. It has half a million stars, at a distance of 34,000 light years. Its diameter is about 100 light years, but most of the stars are in its "core," some 30 light years wide. There are 10,000 times as many stars in this cluster as in any equal space elsewhere in the sky.

Our Galaxy: View from Side and Top, Showing Solar System Position

OUR GALAXY AND OTHERS • The sun, all the visible stars, and billions of stars seen only through a telescope form a huge, flat spiral system known as our galaxy. This great star system is believed to be about 100,000 light years in diameter, but less than 7,000 light years thick at our location. Our sun is close to the equatorial (long) plane of the galaxy, but well off to one side. The galactic center or nucleus appears to be 26,000 light years away, toward Sagittarius (p. 77). Within the galaxy are many star clusters and great clouds of cosmic dust.

The Great Spiral Nebula M31 in Andromeda Mt. Wilson

Our galaxy is rotating like a big whirlpool, and the myriads of stars move around its center somewhat as the planets rotate around our sun. Stars near the center rotate faster than those farther out. The two Magellanic Clouds in the southern hemisphere sky, about 150,000 light years away, are the nearest galaxies. They are satellites of our system. Our galaxy resembles the great spiral nebula M31 in Andromeda—a galaxy some 2 million light years away and twice the size of our galaxy. Millions of other galaxies have been revealed by telescopes.

Photo of Dense Section in Milky Way — Yerkes

THE MILKY WAY forms a huge, irregular circle of stars tilted about 60 degrees to the celestial equator. Even before the structure of our galaxy was known, the great astronomer Herschel proposed that this concentration of stars was due to the galaxy extending farther in space in some directions than in others. It is now clear that in looking at the Milky Way you are looking down the long direction of our galaxy. As you look through a deeper layer of stars, the stars appear more numerous. The Milky Way has both thin and congested spots. In Sagittarius it is at its brightest, but all of it is a wonder to behold.

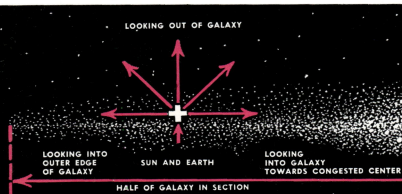

NEBULAE is a term (Latin for "clouds") applied to distant hazy spots in the sky revealed by telescopes. Many are remote galaxies more or less like ours; others are clouds of dust or gas within our galaxy. Of these closer nebulae, the brightest is the Great Nebula in the sword of Orion—diameter 26 light years, distance 1,625 light years. The entire region of Orion has the faint glow of nebulae, but here the glow is strongest. All such nebulae are faint; only long-exposure photographs bring out details of most.

"Horse-head"   Mt. Wilson

Luminous nebulae are found close to bright stars. Short-wave light from these stars stimulates the nebulae to glow like fluorescent lamps. The brightest nebulae are associated with the hottest stars. Low-temperature stars do not cause nebulae to fluoresce, but starlight scattered by the nebulae provides some illumination.

Coal Sack in Crux   Mt. Wilson

Some nebulae, having no stars nearby, are dark. They may obscure bright parts of the Milky Way and be visible as silhouettes. A series of such dark nebulae divides the Milky Way from Cygnus to Scorpius into two parallel bands. Most spectacular of dark nebulae (often called "coal sacks") is the Horse-head Nebula in Orion. Another is in Cygnus near the star Deneb.

Trifid Nebula in Sagittarius Mt. Wilson

Quite odd-looking are the planetary nebulae, so named because they form loose "smoke rings" around stars. The rings look something like the expanding rings around novae, but the latter expand rapidly and disappear relatively soon. Most planetary nebulae are very faint, even as seen through a telescope, but the Ring Nebula in Lyra, the Owl Nebula in Ursa Major, and others are spectacular as they appear in time-exposure photos such as those here.

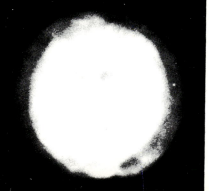

Owl Nebula in Ursa Major Mt. Palomar

Gaseous Nebula in Orion Mt. Wilson

Dust and gas, in the spiral arms of our galaxy, make it hard to determine colors, brightness, and distances of many stars. Dust makes stars behind it look redder, or may obscure them entirely. Bright gas clouds have helped astronomers to trace the spiral arms of our galaxy. Increasingly effective in this work has been the use of radio telescopes, which receive "clues" in the form of radio signals that help to identify chemical compounds in the clouds.

Lick **Ring Nebula in Lyra**

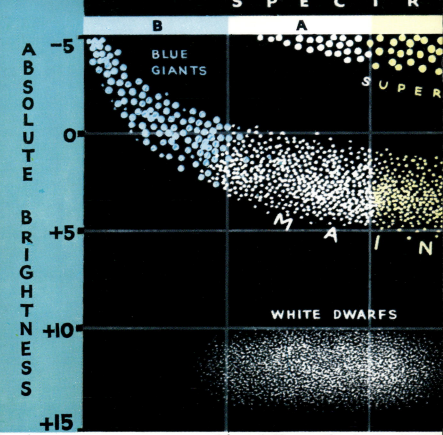

Schematic Diagram Showing Stars in Their

THE ORIGIN OF STARS is a mystery, but the relation between spectral class and absolute brightness of stars, as shown above, offers some clues. Stars evolve out of cosmic gas. As they shrink and grow hotter, they may follow the main sequence, as does our sun. Then, some astronomers think, the stars begin to expand. Hotter ones expand most rapidly; some may shift over into the classes of the giants. Finally, as their substance is exhausted by radiation, stars may flash up in great explosions, becom-

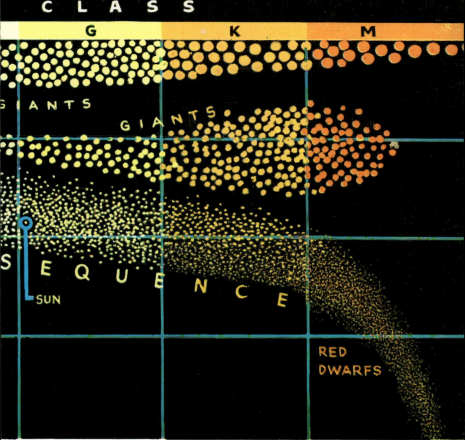

pectral Classes with Absolute Brightness

ing novae, and collapse. They may then become white dwarfs or neutron stars and eventually cease to shine.

Astronomers are perplexed also by evidence suggesting that the entire universe is expanding rapidly. The more distant clusters and galaxies seem to be receding at tremendous speeds. At distances of over a billion light years are "quasars," objects much smaller, yet much brighter, than galaxies. Astronomers are not clear as to their exact nature.

THE CONSTELLATIONS

Stars can be weighed and measured, and their brightness, color, and motions have meaning. Constellations are different. They are but figments of man's imagination—handy inventions to help map the sky. Some of these apparent patterns were known and used by our forefathers for thousands of years. Other minor constellations were invented by 17th-century astronomers. People in different parts of the world imagine the stars represent different shapes and things. Most of the characters and objects of the constellations we use come from the myths of the Greeks and Romans.

Although they seem unchanged for a lifetime, even a century, the constellations are changing; the stars in them are gradually shifting their positions. In any constellation some stars may be farther away than others and unrelated to them. They may differ in direction of movement as well as color or spectral class. Because of precession (p. 53), different stars have been and will become the North Star. Many constellations have been recognized since ancient times. Their boundaries were irregular and often vague until astronomers finally established them definitely by international agreement.

Amateurs nevertheless learn stars most easily by using constellations. Use the charts in this section. Constellations near the north pole are charted on pp. 54-55, and south circumpolar constellations on pp. 98-99. For middle-latitude constellations, you will find a map for each of the four seasons, showing major constellations visible during that season in the north temperate zone.

The dome of the heavens is hard to represent on a flat map. The seasonal maps are designed to show most accurately constellations in middle north latitudes. Distortion

Big Dipper in 20th Century, with Stars Moving in Direction of Arrows.

is greatest near the horizons and in the south. Constellation shapes are truer in the 28 individual constellation pictures in this section. Here they are upright. (For their positions relative to each other, see the seasonal maps.) Polaris is always at about the north celestial pole; where it is not shown, an arrow frequently points toward it.

For a complete list of constellations, see pp. 156-157.

Bright stars have names, often Arabic, and are labeled usually in order of brightness by Greek letters and constellation name, as Alpha Scorpii, brightest star in Scorpius, and Beta Cygni, second brightest in Cygnus. Symbols are used in charts in this book as follows:

★ Stars: 1st magnitude   (brighter than 1.5)
◆   2d magnitude   (1.5 to 2.5)
●   3d magnitude   (2.6 to 3.5)
•   4th and 5th magnitude   (fainter than 3.5)
∴ Star clusters and nebulae

Big Dipper in 100,000 A.D.

CIRCUMPOLAR CONSTELLATIONS

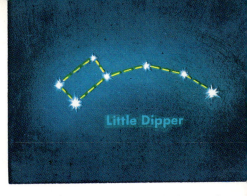
Little Dipper

The term circumpolar constellations implies that the observer is somewhere between the equator and the pole. Each 24 hours, as the earth turns on its axis, the sky seems to wheel overhead. To an observer in the north temperate zone, stars near the pole remain in view as they swing around; stars near the equator rise and set.

At the pole, all constellations are circumpolar; at the equator, none. For places between, the latitude is important, for if a star or constellation is nearer to the pole than the pole is to the horizon, it becomes circumpolar and does not set. The bowl of the Big Dipper does not set at latitude 40 degrees north, but in Florida, at latitude 30 degrees, it does set and so is no longer circumpolar. When the sun is north of the equator during the summer, it becomes a circumpolar star north of the arctic circle. The circumpolar constellations are easy to learn, and you will find them the best place in which to begin your identification of the stars. Once you know the Big Dipper, the rest fall into line. The Little Dipper, Cassiopeia, Cepheus, Draco, and Perseus will guide you to the other constellations shown on the seasonal charts later in this book.

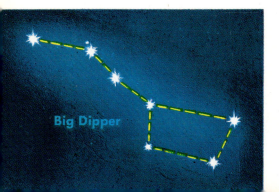
Big Dipper

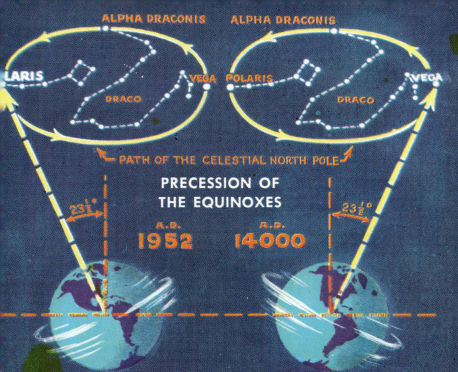

PRECESSION OF THE EQUINOXES

Besides rotating and revolving, the earth has an oscillating motion like that of a spinning top due chiefly to the pull of the moon on Earth's equatorial bulge. Each oscillation takes about 26,000 years. Thus, the North Pole traces a circle on the sky, pointing to different stars as it moves in its circuit. Thus 3,000 years ago Alpha Draconis was the Pole Star. In 14,000 A.D. Vega in Lyra will be the Pole Star and the other constellations will shift accordingly. The Southern Cross will then be visible in the northern hemisphere.

NORTH CIRCUMPOLAR CONSTELLATIONS

At about 40 degrees north latitude the following are considered circumpolar constellations: Big Dipper (Ursa Major); Little Dipper (Ursa Minor); Cassiopeia, the Queen; Cepheus, the King; Draco, the Dragon. To locate these constellations, use the accompanying chart. Facing north, hold the opened book in front of you so that the current month is toward the top. The constellations are now about as you will see them during the current month at 9 p.m. To see how they will appear earlier, turn the chart clockwise; for a later time, counterclockwise. A quarter of a turn will show how much the positions of the stars will change during a six-hour period.

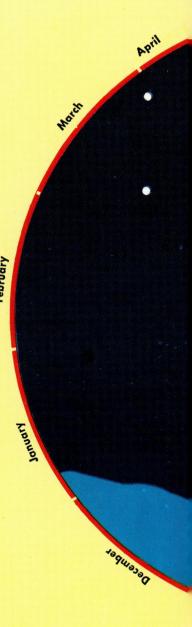

URSA MAJOR, THE GREAT BEAR (BIG DIPPER)

The familiar Dipper is only part of the Great Bear. The Dipper's seven stars are easy to find if you face north on any clear night. The two outer stars of the bowl point to the North Star, Polaris, which is about 30 degrees away. The distance between the pointers is 5 degrees. Both measurements are useful in finding your way around the sky. The middle star of the handle (Mizar) is a double star. Its 4th-magnitude companion is faintly visible, if you look carefully. The rest of the Great Bear spreads as a curve ahead of the pointers and in another curve below the bowl.

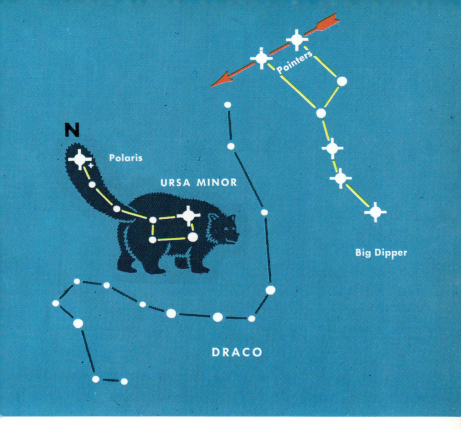

URSA MINOR, THE LITTLE BEAR (LITTLE DIPPER)

Polaris, the Pole Star, is the tail star of the Little Dipper—a dipper which has a reversed curve to the handle. Polaris is a sun, quite like our own, but brighter. Its distance is about 50 light years. Polaris is not exactly at the pole but is less than a degree away; no other 2nd-magnitude star is near it. It is commonly used by navigators to determine latitude. Polaris is a Cepheid variable which changes in magnitude very slightly every four days. The four stars in the bowl of the Little Dipper are of 2nd, 3rd, 4th, and 5th magnitude, making a good scale for judging the brightness of near-by stars.

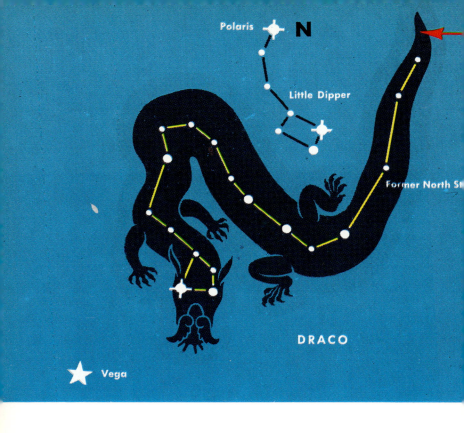

DRACO, THE DRAGON begins, tail first, about 10 degrees from the Big Dipper's pointers, and curves between the two dippers. Then it swings around the Little Dipper, doubles back, and ends in a group of four stars, forming the Dragon's head. These are about 15 degrees from Vega in Lyra. Thuban, in Draco, once the North Star, was the star by which the Egyptians oriented their famous pyramids. Though Draco is circumpolar, it is best seen in late spring and early summer, when it is highest above the northern horizon. A planetary nebula in Draco can be seen with a small telescope.

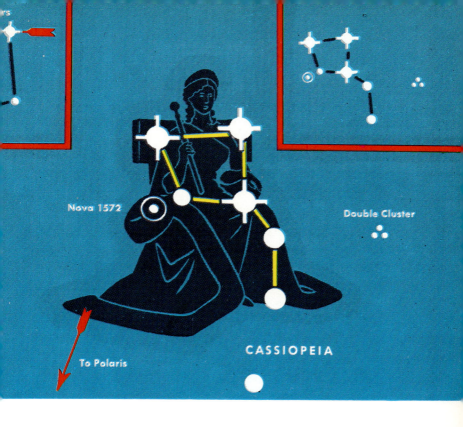

CASSIOPEIA and near-by Cepheus are involved in the myth of Perseus and Andromeda (pp. 80-81). Start at the Big Dipper, at the star where the handle and bowl meet, and sight a line through the Pole Star on to Cassiopeia. Cassiopeia has the shape of either a W, an M, or a chair, depending on how you look at it. Near the spot marked on the map appeared the famous nova of 1572. When this temporary or explosive star appeared, it rapidly increased in magnitude till it was as bright as Venus and could be seen in daylight. Within two years it faded from view.

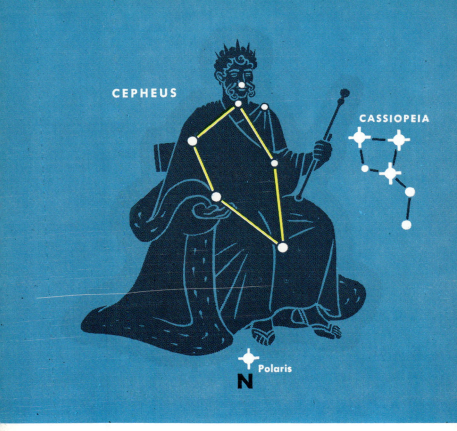

CEPHEUS is closer to the Pole than Cassiopeia. The star, shown above near Cepheus' knee, is about 12 degrees from Polaris on a line with the star forming the end of the W of Cassiopeia. Cepheus is a five-sided figure, like a crude, peaked house. Because of precession, the stars on the west side of Cepheus will successively become the North Star during the next 2,500 to 5,500 years. Just south of the base of Cepheus is a garnet-colored star worth spotting with binoculars. Cepheid variable stars, used in measuring distances of galaxies, are named from Delta Cephei in this constellation (see p. 39).

Circumpolar Stars as a Guide to Key Constellations

A KEY TO CONSTELLATIONS • Use the stars of the Big Dipper and other circumpolar stars to locate one or two important constellations for each season. Follow the curve of the Dipper's handle to Arcturus, or a line through the bottom stars of the Dipper to Gemini. Trace from the end of the handle through the bottom of the bowl to Leo. With these key constellations in mind, you can locate the others more easily. Be sure the constellation you are trying to locate is above the horizon at the season and hour you are looking.

CONSTELLATIONS OF SPRING

The stars of spring, summer, fall, and winter were selected as those easiest to observe at about 9:00 p.m. on the first of April, July, October, and January. The seasons actually begin a week or so earlier.

Each night at the same hour, a star appears slightly to the west of its former position. Hence stars seen in the east at 9:00 p.m. appear higher and higher in the sky at that hour as the season advances. Before April 1, spring constellations are farther to the east, and farther west after that date.

Latitude, as well as season and time of night, determines star positions. The Pole Star's height above the horizon, for example, is the same as your latitude. The seasonal maps are for about 40 degrees north latitude.

In early spring, eleven 1st-magnitude stars are in the sky at once. No other season offers so many. In addition to the constellations on pp. 66-69, look for a number of smaller ones. Between Gemini and Leo lies Cancer, the Crab, a constellation of 4th- and 5th-magnitude stars. At the center of Cancer, note the fuzzy spot. Field glasses or a small telescope brings out details of this open cluster of some 300 stars; it is Praesepe, one of the near-by clusters

in our galaxy. Another larger cluster is Coma Berenices, Berenice's Hair. This is on a line between the tail star of Leo and the end of the Big Dipper's handle. Use field glasses.

In the southern sky are the fainter Corvus, the Crow; Crater, the Cup; and Hydra, the Sea-serpent. Hydra sprawls below Leo and Virgo, the Virgin. It has one 2nd-magnitude star, the reddish Alphard. Corvus, a lop-sided square of 3rd-magnitude stars, is close to Spica in Virgo (see p. 73). Crater is near by, south of Leo. It has one 3rd-magnitude star. South of Corvus is the Southern Cross.

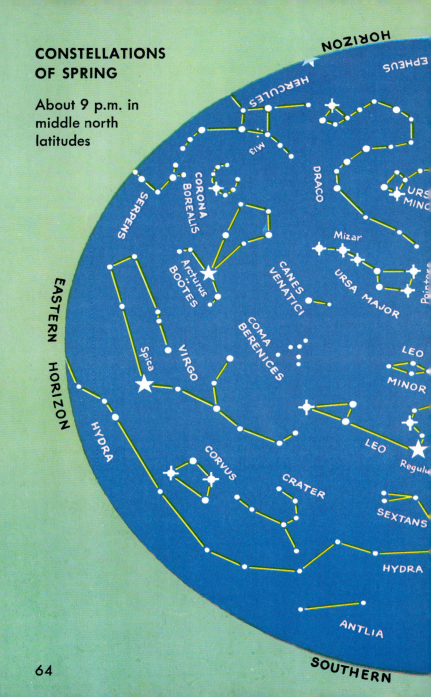

CONSTELLATIONS OF SPRING

About 9 p.m. in middle north latitudes

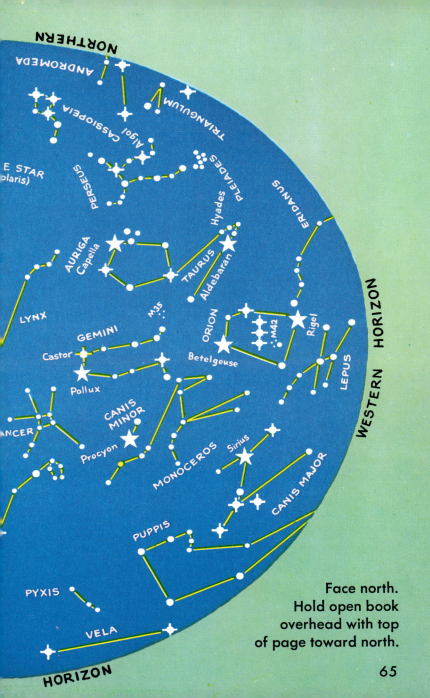

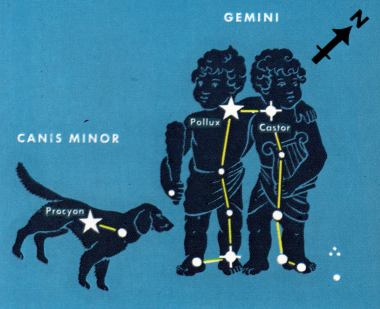

GEMINI, THE TWINS are often considered winter stars, though they are still high in the western sky at the first signs of spring. The bright stars Castor (2nd magnitude, white) and Pollux (1st magnitude, yellow), mark the Twins' heads. They are a scant 5 degrees apart, making good measuring points. Castor is a triple star, and each of its three components is a double star (six in all!). The bottom stars in the Big Dipper's bowl point in the direction of Castor. A line through Rigel and Betelgeuse in Orion points to Pollux. The cluster M35 in Gemini is worth locating with glasses.

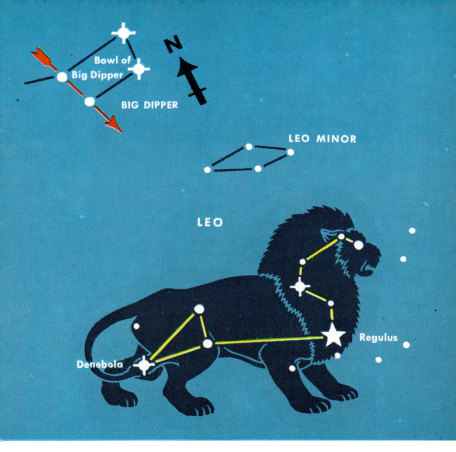

LEO, THE LION is the best known and most conspicuous of the Zodian constellations (pp. 100-101). The Sickle, which clearly forms the Lion's head, is found by following a line through the back stars of the Dipper's bowl southward. Regulus, a blue-white 1st-magnitude star, 86 light years away, marks the base of the Sickle. The pointers of the Big Dipper point in one direction to the North Star; in the other direction, to the triangle that makes up the rear of Leo. The Leonid meteors, a once-spectacular group of "shooting stars," radiate from this part of the sky in mid-November.

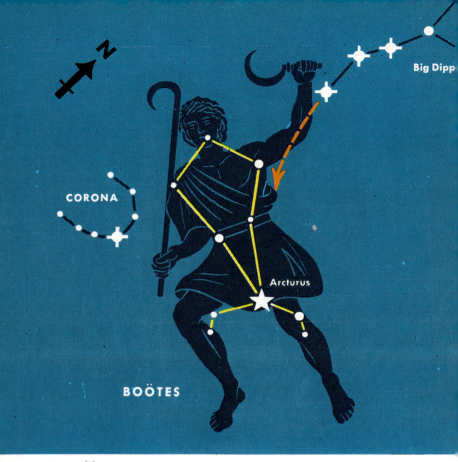

BOÖTES, THE HERDSMAN is found by following the curve of the handle of the Big Dipper 30 degrees to bright, orange Arcturus. The other stars in Boötes are of 3rd and 4th magnitude. Most of them form a kite-shaped figure extending close to the Dipper's handle. Arcturus (magnitude 0.0), one of the few stars mentioned by name in the Bible, is a giant, about 24 times the sun's diameter, 32 light years away. Boötes is chasing the Bears with a pair of Hunting Dogs, which make a small constellation between Arcturus and the Dipper's bowl.

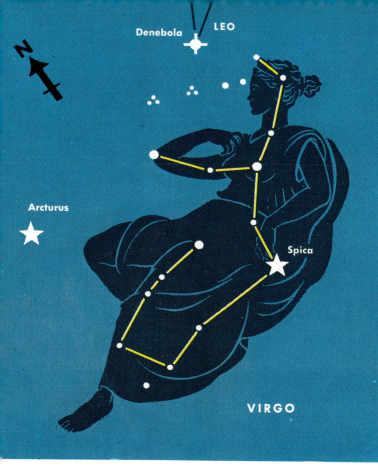

VIRGO, THE VIRGIN begins as a Y-shaped line of stars of 3rd and 4th magnitude extending toward Denebola, tail star of Leo. Spica ends this group; it is a blue-white, 1st-magnitude star 230 light years away. The rest of Virgo is a line of three stars extending on from Spica, and a parallel line of three stars to the north. In Virgo is a cluster of several hundred galaxies about 14 million light years away. A few of the brighter spiral nebulae can be seen with a small telescope. Follow the curve of the Dipper's handle through Arcturus to Spica.

CONSTELLATIONS OF SUMMER

As Leo sinks into the west, a number of new constellations and bright stars move up in the east. Summer is fine for watching them. The weather is likely to be favorable and you may have more leisure. The summer sky is not so brilliant as the early spring sky. You are not likely to see more than six 1st-magnitude stars. However, there are constellations aplenty, and the Milky Way is most impressive in summer.

Boötes, the Herdsman, a late spring constellation, is visible most of the summer, and Arcturus, found by following the curve of the Dipper's handle, is a good place to start exploring the summer sky. Rising just east of Boötes is the Northern Crown, Corona Borealis. Then to the south comes Libra, the Scales, a faint Zodiac constellation. Farther south and east of Libra is Scorpius, the Scorpion, marked by the red, 1st-magnitude star Antares. Moving north again you can trace out the thin line of Serpens, the Serpent, and Ophiuchus, the Serpent-bearer. The two constellations merge. A 2nd-magnitude star marks the head of Ophiuchus, but the remainder of both constellations are 3rd-, 4th-, and 5th-magnitude stars. This double star group has a midway position—midway between the pole and the equator and midway between the points where the sun appears on the first day of spring and fall. Still farther north is Hercules, about due east of the Crown.

East of Hercules is a large triangle of 1st-magnitude stars set on the Milky Way. These are landmarks for the late summer sky, when they are nearly overhead. The three stars are Deneb in Cygnus, the Northern Cross; Vega in Lyra, the Lyre; and Altair in Aquila, the Eagle. Deneb is about 20 degrees east from Vega, and Altair is about 30 degrees from a line between them. South of Altair is Sagittarius, the Archer. Part of it forms a Milk Dipper. Near the triangle are small but bright Delphinus, the Dolphin, and Sagitta, the Arrow. The southern summer sky in  the region of Sagittarius is rich in star clusters and attractive faint stars.

Locating stars is easier if you estimate distances in degrees. A circle contains 360 degrees; the distance from the eastern to the western horizon through the zenith (overhead point) is 180 degrees. From horizon to zenith is 90 degrees. The pointers of the Big Dipper are about 5 degrees apart. From Denebola, at the tail of Leo, to the star at the top of the triangle is 10 degrees. From Rigel to Betelgeuse in Orion is 20 degrees. To avoid confusion as to which direction in the sky is north, which is east, and so on, refer to the pole star. Thus, to find Star A, 16 degrees "south" of Star B, draw an imaginary line from the pole star through Star B; then extend it 16 degrees. East and west are at right angles to this line. West is always the direction of a star's apparent motion as the evening progresses.

Any star can be exactly located on the celestial sphere by using the astronomical equivalent of latitude (declination) and longitude (right ascension). Star atlases are made on this basis. Large telescopes can be quickly directed toward a star whose position is known.

CONSTELLATIONS OF SUMMER

About 9 p.m. in middle north latitudes

Face north. Hold open book overhead, with top of page toward north.

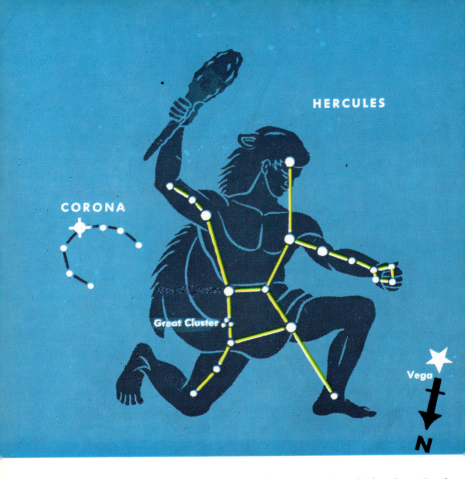

HERCULES lies, upside down, just south of the head of Draco. A line from Arcturus to Deneb in the Northern Cross passes just north of it. A keystone of four 3rd- and 4th-magnitude stars marks the center of Hercules. Along its western edge is the famous cluster M13, 30,000 light years away, with half a million stars. Through a telescope the cluster is a rare sight. The solar system is moving toward Hercules at 12 miles per second. However, because of the rotation of our galaxy, the net movement of the solar system is toward Cygnus.

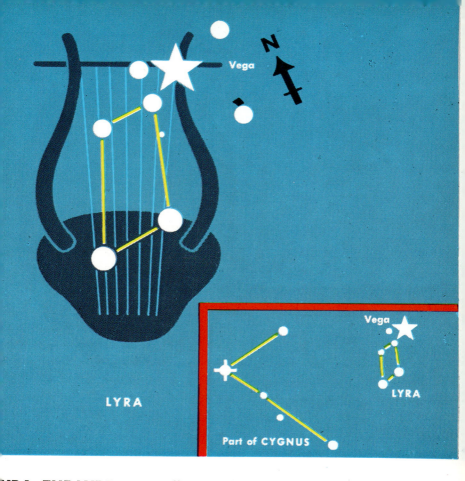

LYRA, THE LYRE is a small constellation marked by the splendor of Vega, its brightest star. Blue-white Vega, magnitude 0.1, 26 light years away, is the brightest summer star. Between the pair of 3rd-magnitude stars at the end of the diamond-shaped constellation is the famed Ring Nebula, 5,400 light years away (pp. 45-47). A large telescope is needed to see its details. One of the 3rd-magnitude stars in Lyra is a binary star. The two stars eclipse every 13 days, the magnitude dropping from 3.4 to 4.3, then rising again.

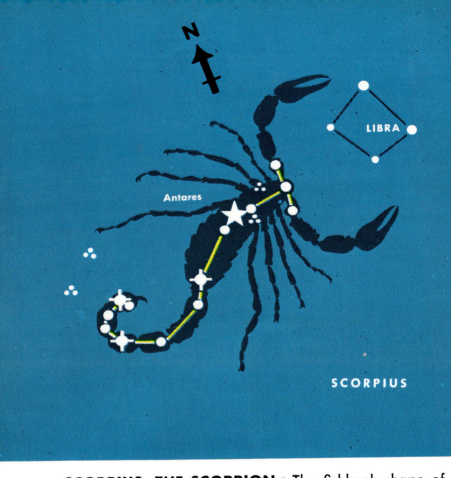

SCORPIUS, THE SCORPION • The fishhook shape of Scorpius in the southern sky is easy to identify. Antares, the red, 1st-magnitude star in the heart of the Scorpion, is a supergiant of a type that gives out much more light than other stars in the same spectral class. Antares' diameter is 390 times that of the sun, but its thin gases have a density of less than one-millionth of the sun's. It is 410 light years away and has a faint green companion. Near Antares and between the tail of Scorpius and Sagittarius are several barely visible star clusters.

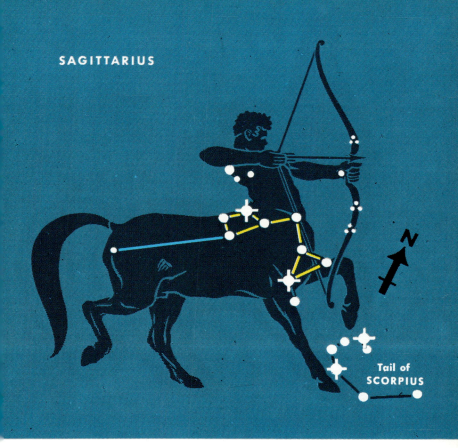

SAGITTARIUS, THE ARCHER lies just east of Scorpius and follows it across the sky. Its central part, called the Milk Dipper, is a small, upside-down dipper. Near the stars of the Archer's bow are several dark nebulae. The general region is rich in star clusters and nebulae. The Milky Way is brightest here. A look at it with glasses or a telescope is exciting. According to myth, Sagittarius is shooting the Scorpion which bit Orion, the Hunter, causing his death. So Orion cannot be seen when Scorpius and Sagittarius are in the sky.

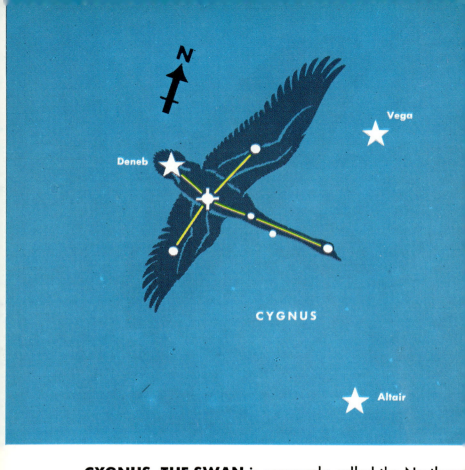

CYGNUS, THE SWAN is commonly called the Northern Cross and actually looks like a cross. Deneb, at the head of the Cross, is in the bright triangle of summer stars. Albireo, a 3rd-magnitude double star, at the head of Cygnus, is almost on a line between Vega and Altair. The Milky Way splits here into parallel streams. The region is rich in varicolored stars—doubles and clusters. It is a region worth exploring. A 5th-magnitude star in Cygnus was the first star measured for distance. It is one of the nearest—10.6 light years away. With Vega and Altair, Deneb in Cygnus makes a conspicuous triangle.

THE EAGLE, THE ARROW, and THE DOLPHIN are three neighboring constellations lying just south of the Cross and the Lyre. Aquila, the Eagle, is a large constellation. Most conspicuous in it is the bright star Altair (magnitude 0.8, distance 16 light years) with a star on either side. The rest of the constellation makes a loose triangle pointed at Sagittarius. Sagitta, the Arrow, an obvious constellation, lies between the Eagle and the Cross. Farther east and forming a triangle with Aquila and Sagitta is Delphinus, the Dolphin, or Job's Coffin, a small diamond of 4th-magnitude stars.

CONSTELLATIONS OF AUTUMN

Some of the autumn constellations follow so closely upon those labeled "summer" that they can be seen well before there is an autumn chill in the air. The autumn constellations are not quite equal to the brilliant skies of spring and summer. The constellation patterns overlap and hence are not so clear. The number of bright stars is limited. But clusters, nebulae, and some unusual stars will entice the observer.

Four of the autumn constellations and two from the circumpolar group (p. 52) are drawn together by a famous Greek legend. No other legend is so well illustrated in constellations as that of the hero Perseus, the winged horse Pegasus, the king and queen of Ethiopia, and their fair daughter Andromeda. King Cepheus and his queen, Cassiopeia (both circumpolar constellations), lived happily till the queen offended the sea nymphs, who sent a sea monster (Cetus) to ravage the coast. The monster would depart only when the royal princess Andromeda was sacrificed. Andromeda was chained to a rock by the sea to await her doom. But just as the sea monster appeared, so did Perseus, the son of Jupiter, flying with winged sandals. Perseus was returning home from a perilous mission. He had just succeeded in killing the dreaded Medusa, a creature with such a terrifying face that mortals who gazed on her turned to stone. (From the blood of Medusa, Pegasus, the winged horse, had sprung.) After bargaining with the king, who promised his daughter to the hero if he saved her, Perseus slew the sea monster. Though the wedding feast was interrupted by a jealous suitor, the pair lived happily thereafter.

All the main characters of the Perseus legend are enshrined as constellations. Cetus, the sea monster, is a spreading constellation of dim stars. The five stars forming the head of Cetus lie in a rough circle southwest of the Pleiades and south of Andromeda. The constellation, extending south and west, has only one 2nd-magnitude star, but also includes a famous variable star, Mira.

Mira is a long-period variable star (p. 39); it was discovered in 1596. Like many other long-period variables, Mira is a red star with a spectrum that fits into Class M. When at its dimmest, Mira is an 8th- to 10th-magnitude star with a temperature of about 3,400 degrees F. Slowly, over a period of around 120 days, its brightness and temperature increase till it is between 2nd and 5th magnitude, and about 4,700 degrees F. Then it slowly dims and cools.

Other minor constellations of autumn include Triangulum, the Triangle, a small group just south of Andromeda between Pegasus and Perseus. About 7 or 8 degrees southwest of the Triangle is Aries, the Ram. Look for a 2nd-, a 3rd-, and a 4th-magnitude star in a 5-degree curve. Pisces, the Fishes, a V-shaped group, fits around the southeast corner of the Square of Pegasus. The Northern Fish is a line of eight 4th- and 5th-magnitude stars. The Western Fish ends in a small circle of 5th- and 6th-magnitude stars just below the Square of Pegasus. Using the two western stars on the Square of Pegasus as pointers, extend a line south nearly 40 degrees and you may see a bright star (magnitude 1.2) close to the southern horizon. This is Fomalhaut, in the constellation of the Southern Fish. The rest of the constellation extends westward as a diamond-shaped group of 4th- and 5th-magnitude stars.

CONSTELLATIONS OF AUTUMN

About 9 p.m. in middle north latitudes

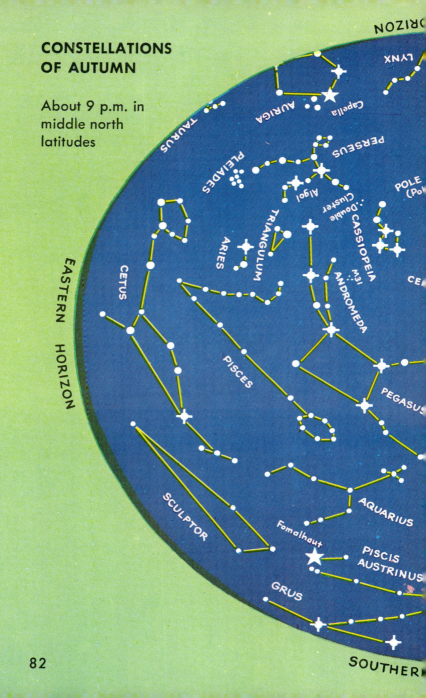

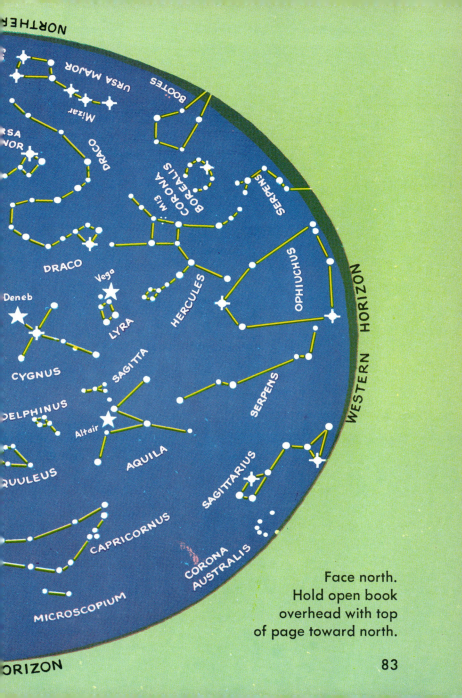

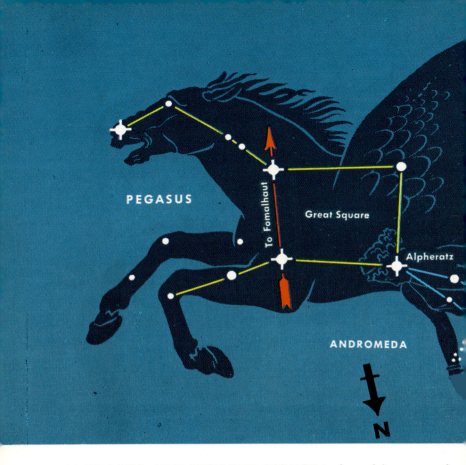

PEGASUS, THE WINGED HORSE is found by extending a line from the Pole Star through the west end of Cassiopeia. This line hits the eastern side of the great square—a rather imperfect square about 15 degrees on each side. West of the square the constellation extends toward Cygnus and Delphinus. Pegasus is upside down, with its head toward the equator. The 2nd-magnitude star Alpheratz, or Alpha Andromedae, is at the point where the constellations join. Pegasus was recorded as a constellation in ancient times.

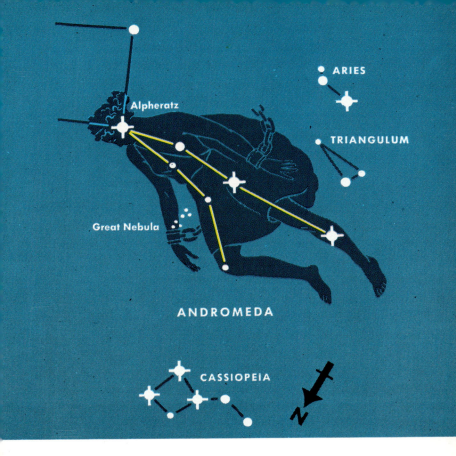

ANDROMEDA in chains extends eastward from Pegasus as two long, spreading lines of stars which meet at Alpheratz, a triple star (2.3, 5.4, and 6.6 magnitude). The northern line of stars extends toward Cassiopeia, the southern to Perseus. In Andromeda is Messier 31, the brightest and largest of the spiral nebulae (pp. 42-43), visible to the naked eye (magnitude 5.0). It is about 2 million light years away and has a diameter of about 150,000 light years. Similar to our own galaxy in many ways, the Messier 31 is about twice as large.

PERSEUS lies close to Cassiopeia. A curved line of stars forming part of Perseus extends toward Auriga. Other stars in Perseus complete its rough, K-shaped figure. The downward side of the K points to the Pleiades. The upward arm ends with Algol, best known of the variable stars—the "Demon Star," or head of Medusa. It is an eclipsing binary. The brighter star is three times the diameter of our sun; the dimmer, even larger. As they revolve, about 13 million miles apart, the dim star eclipses the bright star once every three days, causing a drop from 2.3 magnitude to 3.4.

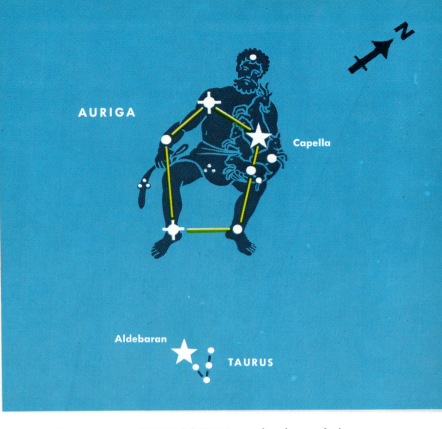

AURIGA, THE CHARIOTEER is the last of the autumn constellations, heralding the coming winter. Auriga lies to the east of Perseus. A line drawn from the top stars of the Big Dipper's bowl points close to Capella, a bright triple star (magnitude 0.1), farthest north of the 1st-magnitude stars. Capella is sometimes known as The Goat; a near-by triangle of stars are the kids. Several open clusters (M37 and M38 especially) are found in Auriga. Each contains about 100 stars and is about 2,700 light years away. The main part of Auriga is a five-sided figure of 1st-, 2nd-, and 3rd-magnitude stars.

CONSTELLATIONS OF WINTER

The sky is never clearer than on cold, sparkling winter nights. It is at those times that the fainter stars are seen in great profusion. Then the careful observer can pick out dim borderline stars and nebulae that cannot be seen when the air is less clear. The winter constellations include some of the brightest and easiest to recognize. Eight 1st-magnitude stars are visible on January evenings, and you may see up to 11 by early spring.

For the watcher of these faint stars the period of accommodation, or getting used to the dark, is important. You will need at least 5 or 10 minutes after looking at a bright light before your eyes will once again see the faintest stars. Use your star map first and then do your observation. If you use a flashlight while observing, cover the lens with red cellophane or with a sheet of thin paper to cut down the intensity of the light. Another good trick in viewing faint stars is to look a bit to one side and not directly at them. This side vision is actually more sensitive than direct vision.

The winter constellations center about Orion, the Great Hunter, who according to the Greek myths boasted that no animal could overcome him. Jupiter sent a scorpion which bit Orion in the heel, killing him. When Orion was placed in the sky, with his two hunting dogs and the hare he was chasing, the scorpion that bit him was placed there too, but on the opposite side of the heavens.

The winter skies also include Taurus, the Bull, of which the Pleiades are a part, and some minor star groups.

Use Orion, so clear and easy to find in the winter sky, as a key to other near-by constellations. The belt of Orion acts as a pointer in two directions. To the northwest, it points toward Aldebaran in Taurus, the Bull, and on, past Aldebaran, toward the Pleiades. In the opposite direction, the belt of Orion points toward Sirius, the Dog Star. Sirius, Procyon (the Little Dog), and Betelgeuse in Orion form a triangle with equal sides about 25 degrees long. South of Orion, about 10 degrees, is Lepus, the Hare; and another 15 degrees south is Columba the Dove (p. 96). A line from Rigel through Betelgeuse points roughly in the direction of Gemini, the Twins.

With such stars as Betelgeuse, Aldebaran, and Rigel in the winter sky, it is worth recalling that these represent an Arabian contribution to astronomy from the 8th to the 12th century. Arabian star names are common. The Greeks, Romans, and their western European descendants gave names to the constellations, most of which represent characters from Greek and Roman myths. Some of the star names are from the Latin, too. Many of the ideas developed by the early astronomers have been discarded, as the limited observations of those days led to incomplete or wrong interpretations. But the names given to stars and constellations have often remained unchanged for centuries and are as useful now as they were long ago.

Because winter nights are long, and often clear, they offer an excellent opportunity for photographing stars and planets. The books listed on p. 11 will tell you more about this interesting hobby. Star trails and photos showing the movement of the moon or other planets can be made with no equipment other than a camera. For other kinds of photographs of heavenly bodies, a polar axis, motor-driven with a clock, is needed to keep the camera accurately following the star or planet.

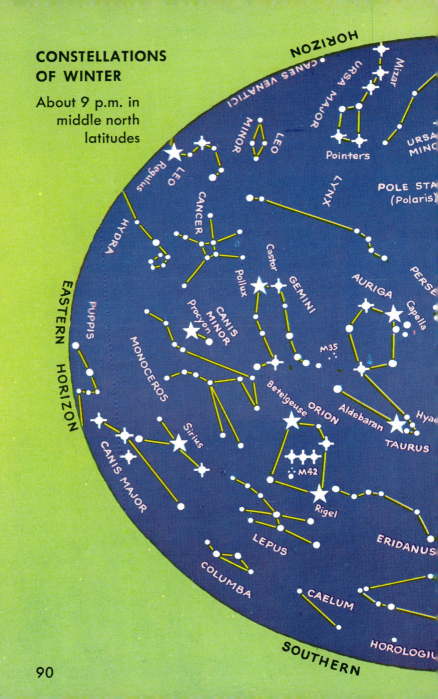

CONSTELLATIONS OF WINTER

About 9 p.m. in middle north latitudes

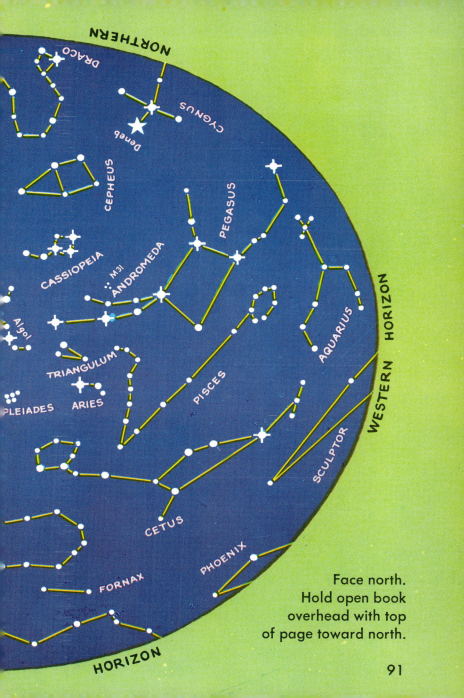

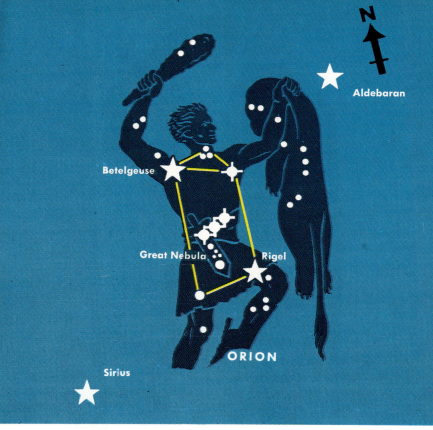

ORION, THE HUNTER is conspicuous and easily remembered. A line drawn from Polaris through Capella in Auriga will bring you to Orion. So will a line from the stars forming the ends of the horns of Taurus, the Bull. The rectangle forming the Hunter's torso is bounded by bright stars. Betelgeuse is a red variable supergiant (magnitude 0.7). Rigel, diagonally opposite but blue-white, is a supergiant double (magnitude 0.1). From Orion's belt, 3 degrees long, hangs the faint sword, containing the great nebula M42, a mass of glowing gas 26 light years in diameter and 1,625 light years away.

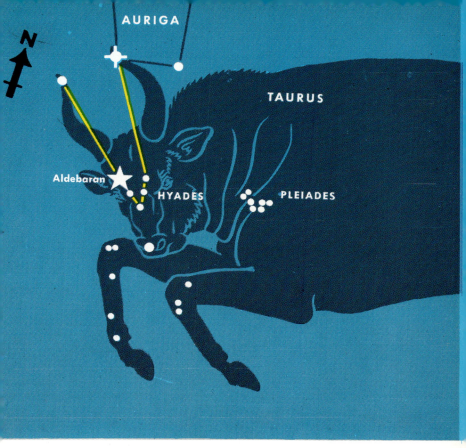

TAURUS, THE BULL represents the form Jupiter took to carry off Europa, a young princess. (Only the forepart of the Bull got into the sky.) Most conspicuous stars in Taurus are the Hyades, which form its face. This clear, V-shaped star group has Aldebaran, a red 1st-magnitude star, at one end. Aldebaran is a double star 68 light years away. The Hyades, actually a loose cluster of about 150 stars, are about 120 light years away. From them extend the horns of Taurus. The 2nd-magnitude star Nath, forming the tip of the Northern Horn, is also part of Auriga and can serve as a guide to Taurus.

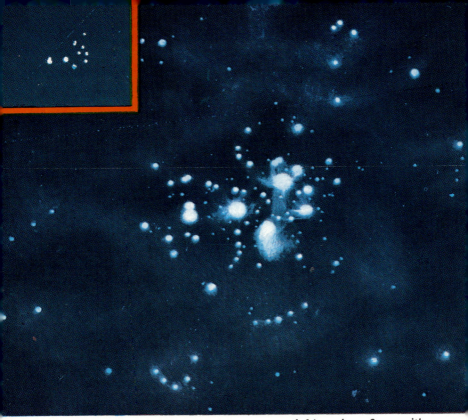

Pleiades as Seen with Unaided Eye (upper left) and as Seen with Binoculars or Small Telescope

THE PLEIADES are a part of Taurus, representing a spot on the Bull's shoulder. They are an open cluster of many stars—at least several hundred—wrapped in a faint nebulosity. Seven stars are visible to the unaided eye. To count them is a test of good eyesight. Field glasses show many more. This cluster and the Hyades are two open clusters in which the stars can be seen without aid. According to the myths, this group of stars represents the seven daughters of Atlas, the giant who supported the world on his shoulders.

CANIS MAJOR AND MINOR are two constellations, each of which has a major star. In Canis Major (the Big Dog) Sirius, brightest of all stars, dominates. With a magnitude of —1.43 it is over 300 times brighter than the faintest visible stars. Sirius, the Dog Star, is only 8.8 light years away. The rest of Canis Major includes double and triple stars and several clusters. Canis Minor, the Little Dog, is smaller and has only one visible star besides Procyon (magnitude 0.4). The belt of Orion points eastward and a little south to Sirius. An eastward line from Betelgeuse in Orion takes you to Procyon.

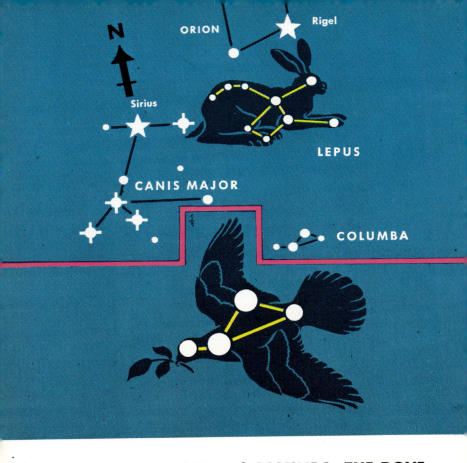

LEPUS, THE HARE and COLUMBA, THE DOVE are small constellations near Orion. Lepus is south of Orion and due west of Canis Major. The main part of the constellation is a four-sided figure of 3rd- and 4th-magnitude stars. Most of the other stars are between this group and Orion. The Dove, which commemorates the dove which flew out from Noah's Ark, is an even smaller constellation south of Lepus and, in most parts of the United States, close to the southern horizon. The four stars form a close group about 5 degrees long.

SOUTHERN HEMISPHERE CONSTELLATIONS

The farther south you go, the more southern stars you can see. At 40 degrees N. latitude about half the southern stars are visible. In southern Florida and Texas the Southern Cross is seen. Southern constellations not thus far described are in a circle (next page) within 40 degrees from the South Pole. They were described first by Magellan and other early observers. Most famous is the Southern Cross, 6 degrees long, pointing to the South Pole. The Centaur, near by, has two 1st-magnitude stars. A companion to one of them is the nearest star to the earth. The Magellanic Clouds (p. 43) are close companions of our galaxy.

Right: Strip of Sky from North Pole to South Pole

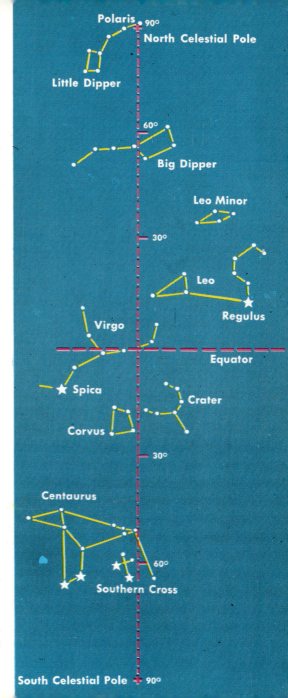

SOUTH CIRCUMPOLAR CONSTELLATIONS

At about 40 degrees south latitude the following are the chief circumpolar constellations: Crux (Southern Cross), Carina (the Keel, of the ship Argo), Volans (Flying Fish), Dorado (Goldfish or Swordfish), Hydrus (Sea Serpent), Tucana (Toucan), Octans (Octant), Pavo (Peacock), Ara (Altar), Triangulum (Southern Triangle), and Centaurus (Centaur).

At about the equator you can locate these constellations with the accompanying chart. Facing south, hold the open book in front of you so that the current month is toward the top. The constellations are now about as you will see them during the current month at 9 p.m. To see how they will appear earlier, turn the chart counterclockwise; for a later time, clockwise. A quarter of a turn will show how much the positions of the stars will change during a six-hour period.

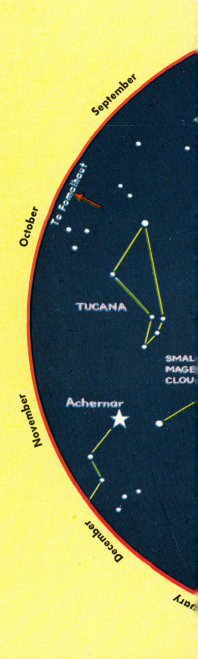

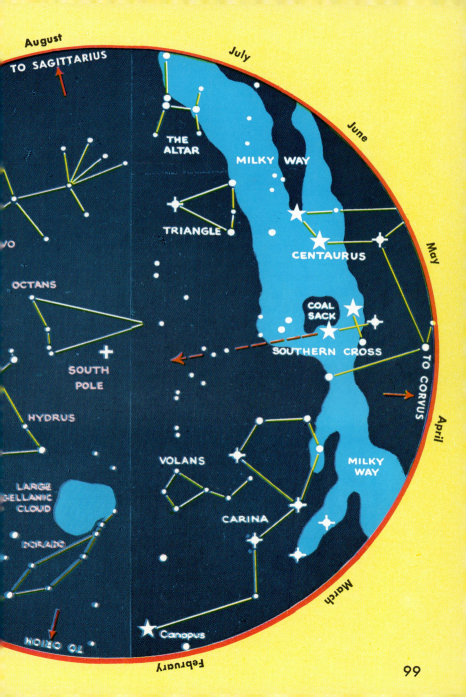

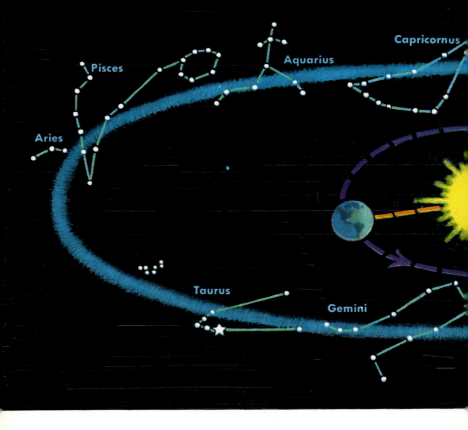

THE ZODIAC is a belt of 12 constellations: Aries, Taurus, Gemini, Cancer, Leo, Virgo, Libra, Scorpius, Sagittarius, Capricornus, Aquarius, Pisces. These star groups circle the sky close to the ecliptic, which is the great circle of the earth's orbit around the sun. The sun, moon, and planets look as though they move against the background of these constellations and seem to be "in" them. Easiest to observe is the moon's path. The journeys of the planets take longer, depending on their distance from the sun.

The sun itself seems to move through the Zodiac constellations each year. The change of constellations seen just before sunrise or after sunset confirms this movement.

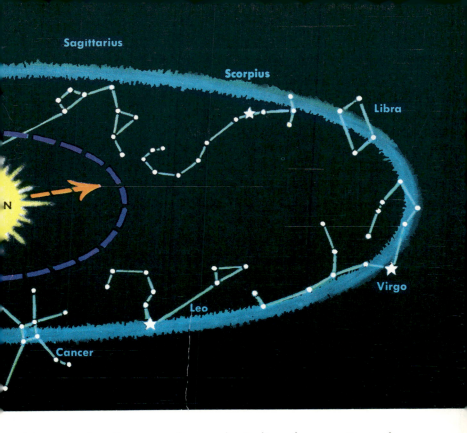

Note, in the diagram, the earth circling the sun. From the earth, the sun seems to be in the constellation Libra. As the earth revolves, the sun will seem to move through Scorpius and Sagittarius, till finally it is back in Libra again.

Babylonians and other ancient astronomers recognized this apparent motion of the sun, moon, and planets. This knowledge helped them predict the seasons. Nowadays the Zodiac is often linked to astrology, which claims to interpret the influence of stars on people and worldly events. Astronomers are convinced that astrology has no scientific foundation.

THE SOLAR SYSTEM

There are altogether in the sun's family 9 planets, 32 moons or satellites, thousands of minor planets or asteroids, scores of comets, and untold millions of meteors. Central star of the solar system, the sun makes up over 99 per cent of its mass (weight).

The planets range from tiny Mercury, which is some 36 million miles from the sun, to farthest Pluto. Mercury goes round the sun in three earth-months. Pluto takes almost 250 earth-years to circle the sun once.

Around some of the planets revolve moons. Although Mercury, Venus, and Pluto have none, the other planets have one to 12. Jupiter, the largest planet, has 12. Of these, 4 can be seen with field glasses or a small telescope.

The rings of Saturn, made of millions of tiny fragments, can be seen in a small telescope as a platelike belt around the planet.

Most asteroids revolve in paths between Mars and Jupiter. The brighter ones can be found by amateurs with telescopes if their positions are known.

Comets, circling the sun in elongated orbits, cruise into view and out again in periods ranging from a few years to several hundred years and much longer.

Meteors, which appear as bright streaks or flashes in the sky, burn because of friction with the earth's atmosphere. Meteors that strike the earth provide material from outer space that one can study at first hand.

Jupiter

Saturn

Uranus

Neptune

Pluto

Only on the earth is life definitely known to exist. Temperatures on the other planets except Mars are probably too extreme to permit plant or animal life as we know it. Mars, as observed by Mariner 9 space-probe in 1972, possesses volcanoes, canyons and just possibly may be inhabited by very simple life forms. Planets outside our solar system may exist.

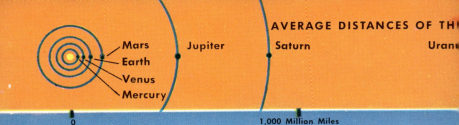

AVERAGE DISTANCES OF TH[E]
Mars, Jupiter, Saturn, Uran[us], Earth, Venus, Mercury

0 — 1,000 Million Miles

THE PLANETS

	Mercury	Venus	Earth
Average distance from sun (in millions of miles)	36	67	93
Distance from sun (compared to earth)	0.39	0.72	1.00
Diameter at equator (in miles)	3,100	7,600	7,913
Mass or weight (compared to earth)	0.05	0.81	1.00
Volume (compared to earth)	0.06	0.92	1.00
Number of moons	0	0	1
Length of day (in hours)	1,416	5,834	24
Length of year (compared to earth)	0.24	0 62	1.00
Inclination of equator to orbit (in degrees)	28	88	23.5
Weight of an object weighing 100 lbs. on earth (in pounds)	25	85	100

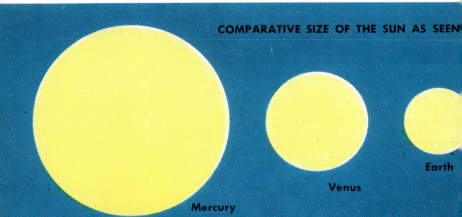

COMPARATIVE SIZE OF THE SUN AS SEEN
Mercury, Venus, Earth

PLANETS FROM THE SUN

Neptune

Pluto

000 Million Miles   3,000 Million Miles

Mars	Jupiter	Saturn	Uranus	Neptune	Pluto
142	483	886	1,782	2,793	3,670
1.52	5.20	9.54	19.19	30.07	39.46
4,140	86,800	71,500	29,400	27,000	3,600 (?)
0.11	318.4	95.3	14.5	17.2	0.2
0.15	1,318	736	64	60	0.1
2	12	10	5	2	0
24.5	10	10	10.7	15.7	154
1.9	12	29	84	165	248
24	3.1	26.8	98	29	(?)
36	264	117	92	112	30 (?)

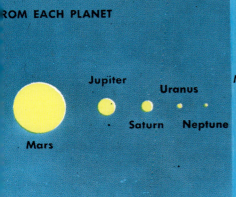

ROM EACH PLANET

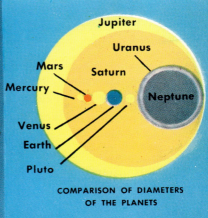

COMPARISON OF DIAMETERS OF THE PLANETS

Laplace Theory of Origin of Solar System

ORIGIN OF THE SOLAR SYSTEM • The earliest and most noteworthy scientific explanation of the origin of the solar system was developed by the French mathematician Pierre de Laplace in 1796. He imagined a globe of hot, glowing gases in space, rotating slowly. As the globe contracted, it rotated faster. The faster rotation caused the globe to flatten out into a disc, which, as it shrank, left rings of gas around it. These rings gradually condensed into planets, some with satellites.

Laplace's theory, called the nebular hypothesis, left many things unexplained and was gradually abandoned. Even today no theory gives a wholly satisfactory explanation of all the observed bodies and motions in the solar system. Some modern astronomers, notably G. P. Kuiper, suggest that as the original solar nebula shrank, it may have left distinct clouds of gas rather than rings around it. These clouds, called "protoplanets," condensed further and became the planets, with their atmospheres and satellites. The satellites, it is believed, could have formed by processes like those that formed the planets.

People often inquire about the possibilities of other solar systems—of stars with planets, and planets supporting life. The number of stars is so great that the statistical possibility of other solar systems does exist. Whatever the series of events that brought our solar system into being, they may have been repeated many times within the tremendous space of the universe. Large planets of a few nearby stars may have been detected, but no system like ours has yet been discovered. Even with the newest telescopes the chances of seeing such a system are slim.

A Modern Theory of Origin of Solar System

Artist's Conception of a Landscape on Airless Mercury

MERCURY, the smallest planet and nearest to the sun, can at times be seen in the east just before sunrise or in the west just after sunset. It has phases like the moon's (p. 140). Mercury orbits the sun every 88 days and rotates on its axis every 58 or 59 days. The direction of rotation is from west to east. The same side of Mercury does not always face the sun, as once believed, and so its temperature is fairly even rather than extremely hot on one side, cold on the other.

Venus in Crescent Phase

VENUS, called the Morning or Evening Star depending on when it is visible, is the nearest of the principal planets—about 26 million miles away at its closest approach. Then it appears through the telescope as a thin crescent. It is brightest a month later—15 times brighter than Sirius, the brightest star. A dense atmosphere conceals the planet's surface, but its surface temperature is far above the boiling point of water. It rotates from east to west.

EARTH, the planet on which we live, gives a basis for understanding the others. The earth has been accurately measured. Its diameter at the equator is 7,926.68 miles; through the poles it is only 7,899.98 miles, or 26.7 miles less. This very slight flattening at the poles leaves the earth an almost perfect sphere. An atmosphere of gases surrounds the earth, extending upward about 500 miles. But the atmosphere decreases rapidly with altitude, becoming thinner and thinner as one goes higher. Half of it is found within 3 miles of the surface. The atmosphere is an essential part of such effects as rainbows, sunrise and sunset colors, and auroras.

The earth has a complex pattern of motions, all of which affect our relationship to the stars and other planets. First, the earth rotates on its axis in four minutes less than 24 hours as measured by your watch. Second, the earth revolves on its 600-million-mile orbit around the sun once a year, at a speed of 18½ miles per second. Third, the earth's axis has a motion, called precession (p. 53), making one turn in about 26,000 years. Fourth, the North and South Poles are not stationary, but wander in rough circles about 40 feet in diameter. Finally, there is a solar motion of 12 miles per second, while our part of the galaxy seems to be whirling through space at 170 miles per second.

The weight of the earth is written as 66 followed by 20 zeros tons. On the average, it is 5½ times as heavy as an equal-sized body of water. However, studies of rock, of earthquake waves, and of gravity show that the earth is not the same throughout. Near the center of the earth the material is under a pressure of about 25,000 tons per square inch. That tremendous pressure creates a very dense core averaging about 10 to 12 times the weight of water. From the center of the earth to the surface is about

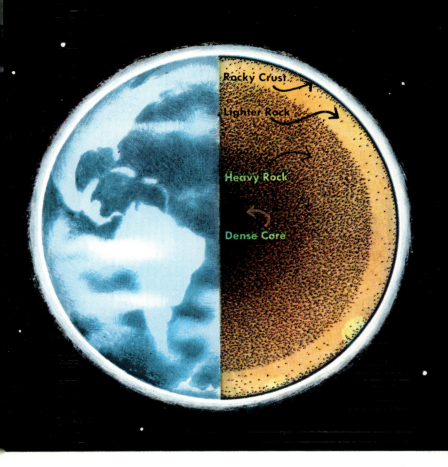

3,950 miles. The first 2,200 is this dense, heavy core of compressed rock. The pressure causes this core to react like a liquid to earthquake waves.

Surrounding the core is a 1,700-mile rigid layer of heavy rock which grades off into lighter rocks nearer the surface. Finally comes the outer rocky crust, up to 25 miles thick, where the density of the rock is only 2.7 times that of water. On the very surface a thin skin of rock has been altered into soil by the action of water and air. It is here that life is concentrated.

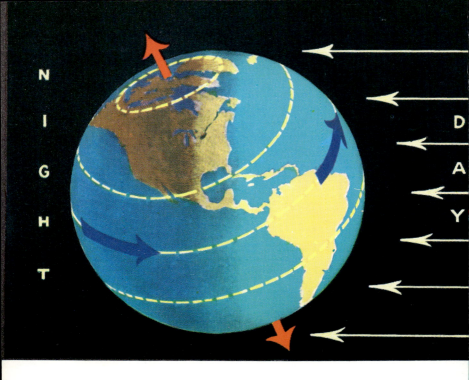

DAY AND NIGHT is due to the earth's rotation. The earth is a dark sphere lighted on one side by the sun. As the earth rotates on its axis from west to east every 24 hours, the sun seems to rise in the east, cross the sky, and set in the west. The earth's atmosphere bends and diffuses the sunlight, before the sun rises, to make dawn, and it keeps the sky light for a time after the sun has set. Day and night are always equal at the equator, but because of the tilting of the earth's axis to the plane of its orbit, only at the first days of spring and fall are day and night equal in middle latitudes. In the northern hemisphere, length of daylight and height of the sun above the horizon at noon increase from the first day of winter to the first day of summer, then decrease again. The polar regions have 24 hour days during summer.

TIME • The earth rotates through 360 degrees in about 24 hours, at the rate of 15 degrees per hour. New York City and Lima, Peru, have the same sun-time, because they have the same longitude. But when it is noon in New York it is still late morning in Chicago.

When the sun reaches its highest point (noon) at a given location, the time at a point 15 degrees west of that location is only 11 o'clock. Local time is therefore different for all places that are not on the same longitude. When only local time was used, New York clocks were 11½ minutes behind Boston, and Washington's were 12½ minutes behind New York. To avoid the confusion resulting from such small differences, in 1883 the nation was divided into time belts, each about 15 degrees wide and each differing by one hour in time. Similar belts now girdle the earth—24 of them.

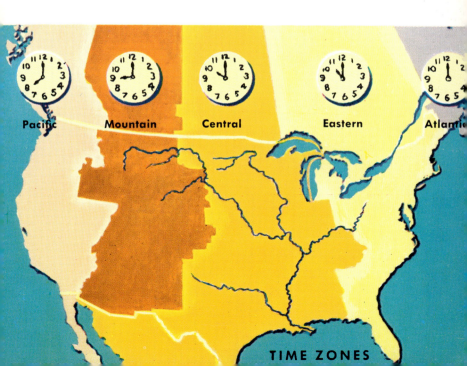

TIME ZONES

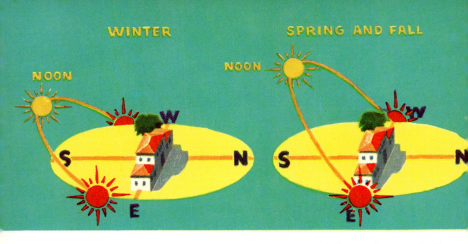

Position of Sun at Sunrise, Noon, and Sunset in Middle Northern Latitudes

EARTH SEASONS • Because of the 23½-degree tilt of the earth's axis, the sun is above the horizon for different lengths of time at different seasons. The tilt determines whether the sun's rays strike us at a low angle or more directly. At New York's latitude the more nearly direct rays on June 22 bring about three times as much heat as the more slanting rays of December 22. Heat received by any region depends on length of daylight and angle of the sun above the horizon. Hence the differences in seasons in different parts of the world.

In the region within 23½ degrees of the poles, the sun remains above the horizon 24 hours a day during some part of the summer. The farther north, the longer it stays above the horizon. Every place above the arctic circle experiences the midnight sun. In the area within 23½ degrees of the equator, the sun is overhead at noon at some time during summer. For latitudes in between, the highest point reached by the sun in summer is 90 degrees minus the latitude, plus 23½ degrees. The low point of the noonday

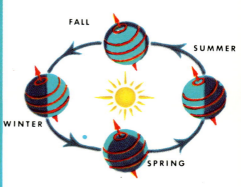

at Beginning of Each Season

sun, in midwinter, is 90 degrees minus the latitude, minus 23½ degrees. At Chicago (latitude 42 degrees north) the height of the noonday sun varies from 24 degrees in winter to 71½ degrees in summer.

Local conditions affect seasonal patterns. Mountain ranges, ocean currents, altitude, prevailing winds, and other factors produce the seasonal climate in a given locality. But the angle of the sun's rays is still a most important factor in determining the plant, animal, and human life in a region.

Positions of Midnight Sun at 15-Minute Intervals above the Arctic Circle

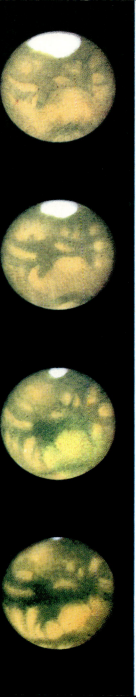
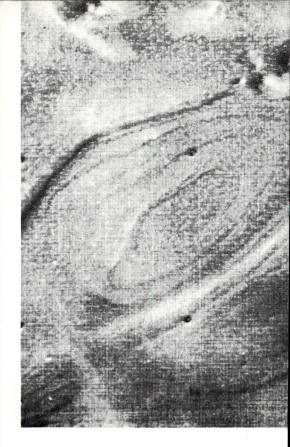

MARS circles our sky in just under two years as it revolves around the sun in some 687 days. It moves in an eccentric orbit; sometimes it is 35 million miles from us, sometimes 234 million miles. The positions best for observation occur every 15 or 17 years (October 1973 was last). Flybys of Mars by Mariner probes showed the surface to be

Shrinkage of a Polar Cap (tilted toward earth) During a 3-Month Period

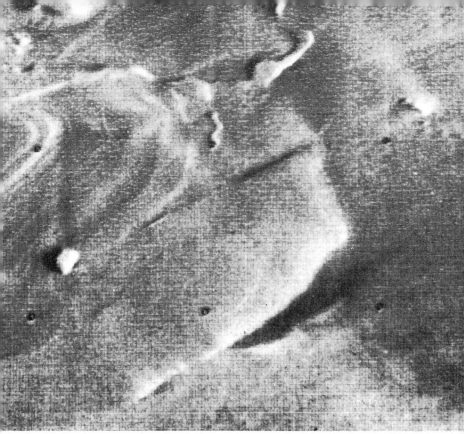

Oval tableland near south pole of Mars; photographed by Mariner 9. (NASA)

covered by craters created by asteroidal impact ranging from several hundred miles to a thousand feet. The surface shows a striking similaritiy to the moon. No canals were found. The reported canals of the past century's observations were made up of unrelated dark spots put together as continuous features by the observer's eye and mind. The polar caps are real and probably are made up of a mixture of snow and dry ice. The atmosphere is largely carbon dioxide. Temperatures may reach 60°F in the day time and —100°F at night.

JUPITER is the largest planet. Never closer than 367 million miles to earth, it takes 12 of our years to circle the Zodiac. Light and dark belts parallel to the equator, in the planet's atmosphere, slowly change. The great red spot, 30,000 miles long, seems more nearly permanent, though fading.

Four of the moons, large and bright, have diameters of 2,300 to 3,200 miles. They revolve around Jupiter in 2 to 17 days and may easily be seen with field glasses. Often one or more are eclipsed by Jupiter or pass before

Jupiter with Red Spot, and Four of Its Twelve Moons

it, throwing small shadows on the clouds. The other eight moons are less than 100 miles in diameter. One, very close to Jupiter, revolves at over 1,000 miles a minute.

Jupiter, fastest-rotating of the planets, turns in less than 10 hours. This speed has produced a pronounced flattening at the poles.

The temperature of Jupiter is close to minus 200 degrees F. Ammonia and methane gases are in the atmosphere, but no water. Ice may exist on its cold, barren surface.

SATURN is a bright "star" to the unaided eye, but its rings can be seen only through a telescope. The closest it gets to the earth in its trip around the sun, which takes 29 of our years, is 745 million miles. It stays two years in each constellation of the Zodiac. Like Jupiter, it is covered with banded clouds. The bands are not so clear as Jupiter's but seem more nearly permanent. Bright spots occasionally appear in them.

The rings were discovered in 1655 by telescope. They are probably made of very many small solid particles.

Saturn with Rings and Six of Its Moons

These may be material which never formed into a satellite, or fragments of a close satellite torn asunder by the tidal pull of Saturn, or ice particles. First is a dull outer ring, next a dark area, then the widest, brightest ring, and inside this a thin, dark space and a dusky "crepe" ring. This inner ring is thin and quite transparent. The rings shine brightly, and when they are tilted toward the earth, Saturn's brightness increases. Outside the rings are ten moons. One, larger than ours, is the only satellite believed to have an atmosphere.

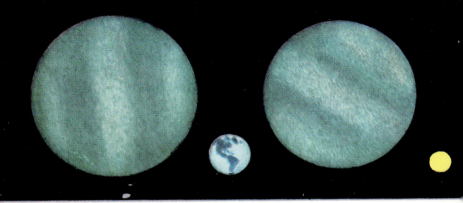

Left to right: Uranus, Earth, Neptune, Pluto

Photographic Plates by Which Pluto Was Discovered (note arrows) Lowel

URANUS, NEPTUNE, and PLUTO • Uranus may be seen by a sharp-eyed observer. The other outer planets are telescopic. Uranus was accidentally discovered in 1781. Its failure to follow its predicted orbit seemed to be due to the gravitational pull of a planet farther out. Two astronomers independently calculated the position of the undiscovered planet, and when telescopes were turned to this region in 1846, Neptune was found. Pluto was discovered in 1930 at the Lowell Observatory as the successful conclusion of a search over many years.

View from Outer Space of Asteroids Between Mars and Jupiter

ASTEROIDS • The 1,700 minor planets now known are all telescopic objects. Most of them were discovered photographically. The largest, Ceres, has a diameter of 480 miles. Most of the rest are less than 50 miles wide. The majority move between the orbits of Mars and Jupiter. Others enter the area between Mars and the sun. Eros, one of these asteroids, will be within 14 million miles of the earth in 1975. Most families of asteroids seem strongly influenced by Jupiter, and move in orbits determined by the gravitational pull of Jupiter and the sun.

LOCATING THE VISIBLE PLANETS

The tables on these pages will help you locate the best-known planets in the constellations where they will be found at various times. The constellations are those of the Zodiac (pp. 100-101). Positions indicated are approximate. To be sure of not mistaking a star for a planet, check appropriate constellation charts if necessary.

For positions of Mercury, Uranus, Neptune, and Pluto, refer to *American Ephemeris and Nautical Almanac* or other yearly astronomical handbooks, or to *Sky and Telescope* magazine.

Italic type indicates morning star; regular type indicates evening star. Dashes indicate planet is too near Sun for observation. (Source: Planet Tables by Fred L. Whipple)

	JANUARY	APRIL	JULY	OCTOBER
		VENUS		
1975	Capricornus	Taurus	Leo	Leo
1976	*Ophiuchus*	*Pisces*	—	Libra
1977	Aquarius	—	*Taurus*	Virgo
1978	—	Aries	Leo	Libra
1979	*Scorpius*	Aquarius	—	—
1980	Capricornus	Taurus	Taurus	Leo

Jupiter in Pisces, July 1975

Path of Saturn 1970—1985

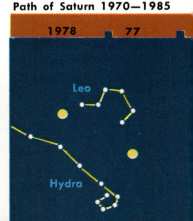

Venus in Leo, July 1978 **Mars in Gemini, April 1976**

	MARS			
1975	Sagittarius	Aquarius	Aries	Taurus
1976	Taurus	Gemini	Leo	Virgo
1977	Sagittarius	Aquarius	Taurus	Gemini
1978	Cancer	Cancer	Leo	Libra
1979	—	Pisces	Taurus	Cancer
1980	Leo	Leo	Virgo	Scorpius
	JUPITER			
1975	Aquarius	—	Pisces	Pisces
1976	Pisces	—	Taurus	Taurus
1977	Aries	Taurus	Taurus	Gemini
1978	Taurus	Gemini	Gemini	Cancer
1979	Cancer	Cancer	Cancer	Leo
1980	Leo	Leo	Leo	Virgo
	SATURN			
1975	Gemini	Gemini	—	Cancer
1976	Cancer	Gemini	—	Cancer
1977	Cancer	Cancer	Cancer	Leo
1978	Leo	Leo	Leo	Leo
1979	Leo	Leo	Leo	Leo
1980	Virgo	Leo	Virgo	Virgo

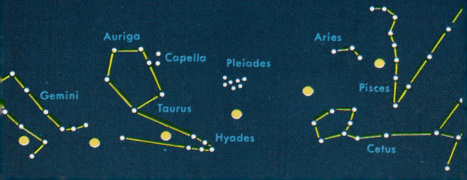

Path of Halley's Comet

COMETS appear five to ten times yearly. Most are faint telescopic objects, but the chance of seeing at least one large, bright comet during your life is good. Records of comets, which often caused terror in ancient times, go back at least 25 centuries. The famous Halley's comet was reported as long ago as 240 B.C.

Comets are among the smaller members of the solar system, moving around the sun in definite orbits. Most have orbital periods of tens of thousands of years. Comets seem to be closely related to meteors and perhaps to asteroids. The head of a comet may consist of ice and meteoric particles. Many comets and meteor swarms have similar orbits. A few comets have not reappeared as

Path of Encke's Comet

Halley's Comet, Morning of May 13, 1910

predicted, but instead, meteor showers have occurred. Comets shine partly by reflected light, partly because sunlight causes their own gases to glow.

Comets' tails consist of exceedingly fine gases and dust, expelled from the heads of the comets. The tail may be so tenuous that stars shine through it with undiminished brightness. (The earth has passed through the tail of Halley's comet with no apparent effect.) A comet's tail always streams away from the sun, and as the comet recedes from the sun the size of the tail decreases. Most comets lack a spectacular tail; in others it develops rapidly and may become hundreds of millions of miles long.

Effect of Sun's Rays on Tail of Comet

Above: Great Comet of 1861

Morehouse's Comet   **Eclipse Comet of 1948**

SOME FAMOUS COMETS

Halley's—Last seen in 1910. Next expected about 1986. The only conspicuous comet that returns in less than 100 years. It has been observed on 28 returns.

Sept. 1882—A brilliant comet, disrupted after passing close to the sun. Split into four comets, which may return between 2500 and 2800.

Biela's—In 1846 this comet split. In 1852, twin comets returned on the same orbit. In 1872 and in 1885, the comets failed to return but brilliant meteor showers were seen instead.

Encke's—A small telescopic comet of Jupiter's family. It has returned regularly since 1819 at 3.3-year intervals.

Schwassmann-Wachmann—This comet, discovered in 1925, has an orbit which lies entirely between the orbits of Jupiter and Saturn, giving it the path of a planet.

Left: Great Comet of 1843

METEORS are generally stony or metallic particles which become separately visible when they plunge into our atmosphere. Though 100 million or more strike the earth's atmosphere daily, those larger than dust particles vaporize. Occasional larger pieces penetrate the atmosphere and strike the earth.

Meteor fragments that reach the ground are known as meteorites. They vary from bits hardly larger than dust particles to chunks weighing tons. The average meteor is estimated to weigh 0.0005 ounce. The two largest known meteorites were found in Southwest Africa (Hoba meteorite) and Greenland (Ahnighito meteorite). Both are from the nickel-iron type of meteor. Another type, the stony meteor, is smaller or is broken up more in falling. No stony meteorites larger than about a ton have been found. Near Winslow in Arizona, east of Hudson Bay, and in Siberia and Esthonia are large craters made when giant meteors struck the earth. Small meteorites have been found near the Arizona crater, but the giant one, estimated at over 50,000 tons, has never been discovered. It probably vaporized on its impact.

Meteorites are often covered with a smooth black crust formed as the tremendous heat caused by the friction of the air fused the meteor's surface. Inside, the

IMPORTANT ANNUAL METEOR SHOWERS

Date	Shower		Location of Radiant
Jan. 2-3	Quadrantids	E	Between Boötes and head of Draco
Apr. 20-22	Lyrids	NE	Between Vega and Hercules
May 4-6	Aquarids	E	SW of the Square of Pegasus
Aug. 10-13	Perseids	NE	Perseus
Oct. 8-10	Draconids	E	Brilliant in 1946. Period about 6½ years
Oct. 18-23	Orionids	E	Between Orion and Gemini
Nov. 8-10	Taurids	NE	Between Taurus, Auriga, and Perseus
Dec. 10-12	Geminids	E	Near Castor in Gemini

Large Iron Meteorite from Greenland: 34 Tons

Etched Surface of an Iron Meteorite

Meteor Crater, Arizona

meteorite either is stony or is composed of iron-nickel alloys, which usually show peculiar crystal patterns. In all, about 30 chemical elements have been found in meteorites.

Meteors may be seen on almost any clear night, though they are more common in the hours after midnight. An observer can usually see about 10 per hour. Occasionally great meteor showers fill the sky with celestial fireworks. These are rare, but lesser showers of meteors can be seen periodically during the year. Some 600 "streams" of meteors are believed to exist, but only a few provide spectacular showers that the amateur can count on seeing.

Stony Meteorite

HOW TO OBSERVE METEORS

Amateurs who know the stars and constellations can study and map meteor showers. Most are best seen between midnight and dawn on nights when showers are expected. Observation works best when two or more amateurs co-operate. Several observers facing in different directions can thus cover the entire sky, and few meteors escape notice. When possible, each observer should have an assistant to record the data, because meteors may sometimes come so fast that one cannot take time out to record them. For each meteor observed it is important to record the time; the star and constellation near which it is first seen; its direction; and the length of its path, in degrees. Notes on its speed (meteors vary considerably), color, trail, and brilliance as compared to stars of known magnitude are also worth while.

Prepare data sheets in advance. Develop abbreviations for quick recording. Find a place with a clear view of the sky and arrange for deck chairs or some other comfortable rest. Warm clothing and a blanket, even in summer, are advisable. Amateurs have recorded hundreds of meteors in an hour of observation. Each, plotted on a chart of the sky, gives a picture similar to that on the next page, clearly indicating the radiant point of the shower.

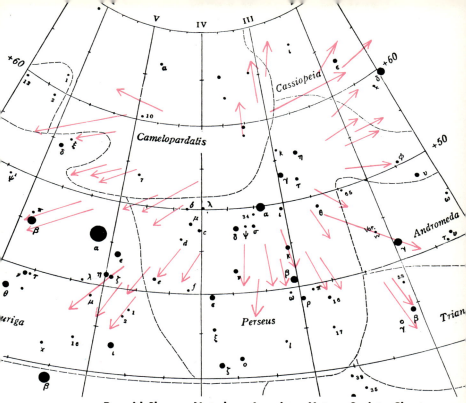

Perseid Shower Noted on American Meteor Society Chart

METEOR CHART • This is a chart of the paths that meteors seem to follow during a shower. Actually, the meteors move in parallel paths. These paths seem to emerge from a point—an optical illusion due to perspective. Meteors in a shower do originate in the same part of the sky, though they are not related to the constellation from which they seem to come. If the radiant point and the speed of the meteors are known, the orbit of the meteor swarm can be calculated. The more observations, the more accurately this can be done. Possibly most meteors belong to swarms and very few solitary, stray meteors exist.

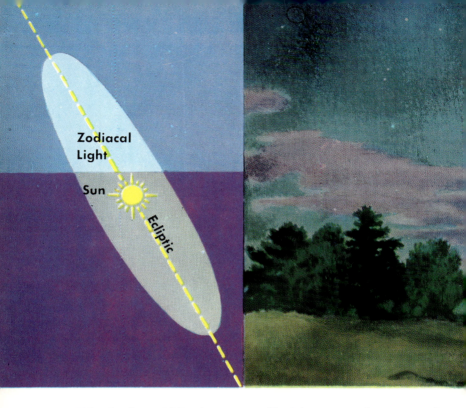

ZODIACAL LIGHT is so called because the triangular band of light which extends from the horizon half way to the zenith follows the earth's ecliptic and hence passes through the constellations of the zodiac. This faint glowing light is best observed in the early evenings of March and April and just before dawn in September and October. In the tropics it is seen more often. On a clear, moonless night its brightest areas may outshine the Milky Way. In the above illustrations its brightness has been emphasized to show its form.

Zodiacal light is apparently sunlight reflected from meteoric particles existing in areas near the plane of the ecliptic. Though meteoric particles are concentrated in

this region, they are widely separated. If the particles were of pinhead size and five miles apart, there would be enough within the earth's orbit to reflect the amount of light usually observed.

The zodiacal light seems to widen into a spot some 10 degrees in diameter at a place just opposite the sun. This faint haze of light that moves opposite the sun is known as Gegenschein or Counterglow.

The area of zodiacal light called Gegenschein may owe its increased light to the fact that meteoric particles directly opposite the sun reflect toward us more sunlight than is reflected by particles in portions of the band that are not directly opposite the sun.

ON THE MOON • Our unique moon is over a quarter of the diameter of the planet around which it revolves. However, its weight is only 1/83 that of the earth, its volume 1/50, and its gravitational pull 1/5 of the earth's. No life forms have been found there. On the sunny side, temperatures are near boiling; on the dark side they are lower than any on earth. In some sections cindery, dusty plains extend in all directions, their surface marred by deep cracks and broken ridges. Thousands of craters, some caused by meteors, some perhaps by ancient volca-

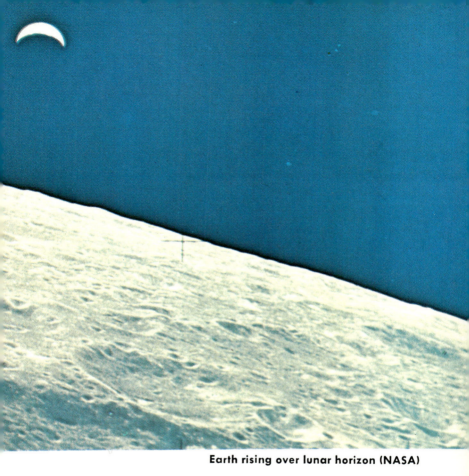
Earth rising over lunar horizon (NASA)

noes, cover the rest of the moon's surface. These range from ¼ mile to 150 miles across, with steep, rocky walls jutting upward as high as a mile or two. Sometimes an isolated peak is within the crater. Bright streaks or rays extend in all directions from some craters.

Besides the craters and plains (called seas by early astronomers who thought they were full of water), the moon has mountain ranges with peaks three, four, and five miles high. In proportion to the size of the moon, they are much higher than mountains on the earth.

ROCKET TO THE MOON • When the first artificial earth satellites were successful (Soviet, October 4, 1957; U.S. February 1, 1958), experts predicted that rockets capable of reaching the moon would soon be developed. Speeds of about 7 miles per second, or 25,000 miles per hour, are required to carry rockets past the point where the earth's gravitational pull drags them back. Once that point is reached no fuel is needed unless special maneuvers or return of a manned vehicle is planned. The rocket coasts at full speed through space, no longer slowed down by atmospheric friction.

In early 1959, such space probes were launched in orbit of the sun and by late fall a probe had landed on the moon and another had transmitted pictures of the moon's far side. Two astronauts, Neil Armstrong and Edwin E. Aldrin, Jr., first landed on the moon on July 20, 1969 (Apollo 11), while a third, Mike Collins orbited the moon. Wearing space suits, they explored a small area near the landing craft on foot. During a landing in 1971 (Apollo 15) a battery-powered vehicle was used to drive about the lunar surface, which allowed for more extensive exploration. On each landing scientific instruments were left behind to record data, such as seismic events, magnetism, variations in temperature, and other lunar conditions, and to transmit it back to earth. Samples of lunar dust and rock were collected by the astronauts and returned to earth for scientific analyses. Manned flights to Mars and Venus are being planned.

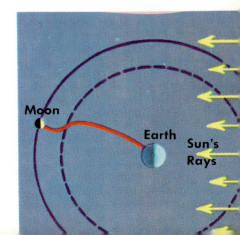

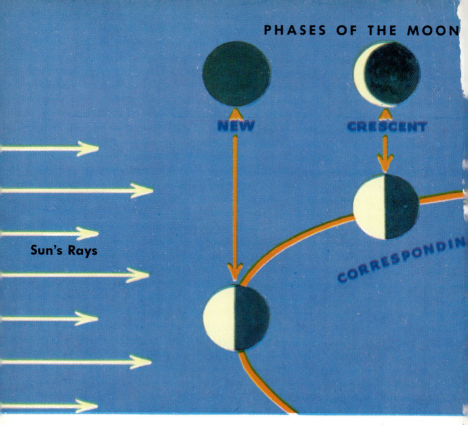

PHASES OF THE MOON • On the earth we see the moon change from crescent to full and back again in 29½ days. If you were out in space, you would see that about half the moon is always lit up by the sun and half is always in darkness, except during an eclipse. When the moon is most directly between us and the sun, we see only the dark side. But, because the moon is revolving around the earth every 27⅓ days, varying amounts of the lighter side are seen.

When the earth is in line between the sun and the moon, we see the moon's fully illuminated side and can watch it as a full moon from sunset to the next sunrise.

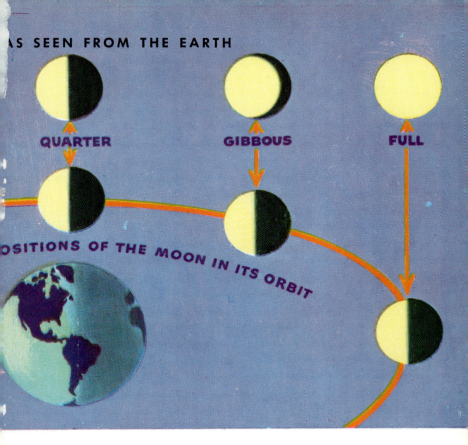

All other stages are in between. When the moon is a quarter of the way around its orbit, we still see half its surface, but half of this half is dark and half is illuminated, giving us a quarter moon. When the moon is more than a quarter, but less than full, it is called "gibbous."

As the moon revolves around the earth, it rotates on its axis, keeping almost the same face turned toward us. A slight wobble of the moon, due mainly to a small difference between the times of revolution and rotation and to a tilt of the moon's axis, has allowed us to see from the earth a total of 59 per cent of the moon's surface. Lunar probes have now mapped the entire surface.

EARTHSHINE is light which has traveled from the sun to the earth, hence to the moon, and back again to the earth. Earthshine is very faint because only a small part of the sunlight reflected from the earth hits the moon. The moon reflects only 7 per cent of this into space again and only a tiny fraction of this 7 per cent finally comes back to earth. When you look for earthshine, note that the crescent moon, because it is brighter, seems larger.

LUNAR HALO • Rings around the sun and moon are often seen. These are in our atmosphere and are of the same general nature as rainbows. Halos are due to the refraction of sunlight or moonlight by thin, high, icy clouds. The halo making a 22-degree circle around the moon is the most common. A 46-degree circle may also form and, if the ice crystals in the clouds are just right, one may see arcs and other curious effects. Halos are usually colorless but sometimes they appear like faint rainbows with the red on the inside.

TIDES • The gravitational pull of a planet, satellite, or star decreases with distance and with smaller weight or mass. The moon, though small, has a strong gravitational pull on the earth because it is relatively very near. The sun has a strong pull because of its great weight. This gravitational pull holds both the earth and the moon in their orbits. The pulls of the sun and the moon cause the tides, the moon having the greater influence.

The pull of the moon is greater on the side of the earth nearest it, and is less on the opposite side of the earth, which is farther away. The difference between the lesser and the stronger pull is equal to a pull away from the moon. These pulls in opposite directions cause the oceans to flow toward the axis of the pulling. The result is a bulge of several feet or more in the oceans on opposite sides of the earth.

As the earth rotates and the moon revolves, the bulge of tides moves also, giving most places alternating high and low tides. The tidal pull of the more distant sun is about half that of the moon. When both pull in line at new and full moon, the tides are higher. These are known

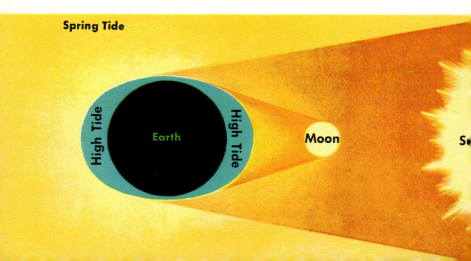

Neap Tide

as spring tides. When they pull at right angles and partly counteract one another, the tides are lower and are called neap tides.

Tide schedules ranging far into the future can be prepared on the basis of the moon's predicted movements. The shapes of ocean basins and seacoasts may determine actual heights of local tides. In general the tidal range is 3 to 10 feet. In V-shaped bays, as the Bay of Fundy, the tide may rise 30 to 50 feet. In broad bays it may rise a foot or less.

Photograph of Moon Taken Through the 36-inch Refracting Telescope at the Lick Observatory

EXPLORING THE MOON • The accompanying chart shows the moon as it might be seen through a low-power telescope. The majority of the various features mapped at this magnification can be seen also with a pair of good, high-power field glasses. Most "craters" and mountains on the moon have the names (often Latinized) of famous astronomers and other scientists. The dry "seas" have fanciful Latin names. Though the chart shows the full moon, studying it as suggested on p. 148 is most satisfactory.

KEY: Mountains, seas, and lakes are in SMALL CAPITALS
Craters and other features are in regular type

Piccolomi
MARE NECTARIS
Vendelinus
Langrenus
Theophilus
PYRENEES MTS.
MARE FOECUNDITATIS
MARE TRANQUILLITATI
MARE CRISIUM
Posidonius
LACUS SOMNIORU

OBSERVING THE MOON • With the photograph and map on pages 146-147 as a guide, you can easily study the moon and identify a dozen or two of the most prominent features. Even a pair of small field glasses will show the "seas," mountain ranges, and ringed plains, and the great "craters." Larger field glasses or a small telescope will disclose all the features identified on the map.

The very best time to observe the moon is in the two- or three-day period after the first quarter. The moon is then in a good position for evening study; nearly all major features can be seen and the moon is not sufficiently bright to cause loss of detail through glare. It is even better to follow the moon evening after evening from its first thin crescent till it is full. As the line of darkness recedes, features near the border stand out in bold relief; the shadows become stronger and details are more easily seen.

Should you pass this stage and desire to explore the moon further, more detailed maps are available. These divide the full moon up into sections and show you the features of each section. Over 500 features on the moon have been named. Its whole surface has been mapped. Detailed study will call your attention to interesting problems: the distribution of craters; the overlapping of some craters; the nature of rays on the moon; and the origin of the seas. Even though the moon is our nearest neighbor, there is much we have yet to learn about it. However, since we have visited the moon and brought back rocks from it, the newer exploration will probably be chemical rather than astronomical.

Total Eclipse of the Sun Showing Prominences and Outer Corona

ECLIPSE OF THE SUN • Until recently no astronomical event offered such opportunities as a total eclipse of the sun. Using new methods, astronomers can now make some of the observations that once had to await an eclipse. But the beauty and awesomeness of a total eclipse are still unequaled. In the pattern of movements of moon, earth, and sun, there are always two to five solar eclipses each year. Some are total, some partial, and some annular (p. 152). On the average there are two total eclipses of the sun every three years.

Total Eclipse of Sun

SOLAR ECLIPSE • An eclipse of the sun can occur only when the moon is new—when it is between the earth and the sun. If the orbits of earth and moon were on exactly the same plane and if these two bodies were at their minimum distance apart, an eclipse would occur every month. It does not occur that often because the moon's orbit is inclined about 5 degrees to the earth's orbit. In addition, the moon's path takes it slightly nearer or farther from earth as it revolves. This is important, as the average length of the moon's shadow is 232,000 miles, but its distance from earth averages 235,000 miles. A total eclipse cannot occur under average conditions.

However, because of variations in its orbit, the moon's shadow is sometimes longer and its distance from earth sometimes shorter. If this occurs at the time of a new moon, we may have an eclipse. The time and place of earth eclipses are calculated years in advance. At any one place the duration of totality varies. In the eclipse of

June 30, 1973, the maximum totality was about 7 minutes, slighlty less than the one on June 22, 1955.

You will see a total eclipse when the true shadow (umbra) of the moon passes over you. The umbra produces a round shadow, never more than 170 miles in diameter, which travels rapidly over the earth. The penumbra, which surrounds the umbra like an inverted cone, does not completely exclude the sunlight and hence gives only a partial eclipse. It forms a circle about 4,000 miles in diameter around the umbra. Observers in the path of totality see a partial eclipse as the disc of the moon covers more and more of the sun's face. Then, at the moment of totality, red prominences appear. The weirdly darkened sky is lit up by the streaming corona, which may extend over a million miles from the sun's surface. Nothing is as inspiring and awesome as the few minutes of totality. Then, after repeating the partial phase, the eclipse is over.

ANNULAR ECLIPSE • The distance from the earth to the moon varies. If an eclipse occurs when the moon is its average distance away or farther, the umbra of the moon's shadow does not reach the earth. An annulus, or thin ring of sunlight, remains around the moon. The path of an annular eclipse is about 30 miles wider than that of a total eclipse. Surrounding this area, as in a total eclipse, is a region 4,000 to 6,000 miles wide where the eclipse is partial. Of all solar eclipses, about 35 per cent are partial; 32 per cent annular; 5 per cent both annular and total; and 28 per cent total.

Right: Total Eclipses Visible in the Northern Hemisphere, 1952-1986

Left, Annular Eclipse. Red Spot Shows Area in Which Annular Eclipse Is Seen; Elsewhere in the Penumbra Eclipse Is Partial

TOTAL ECLIPSES OF SUN • The movements of sun, moon, and earth causing eclipses are well known. They occur in a cycle of just over 18 years, after which a new series of eclipses repeats with only minor changes. One change is a westward shift with each new cycle. A knowledge of these cycles enables astronomers to predict eclipses hundreds of years in advance. Ancient eclipses are the most certain and useful of chronological data.

The next total eclipse of the sun visible in the United States will take place on February 26, 1979 and will be seen in the Pacific Northwest. Not until August 21, 2017 will another total eclipse be visible from the United States. Annular eclipses will be seen in the United States on May 30, 1984 and May 10, 1994. During a solar eclipse, totality may last as long as seven minutes.

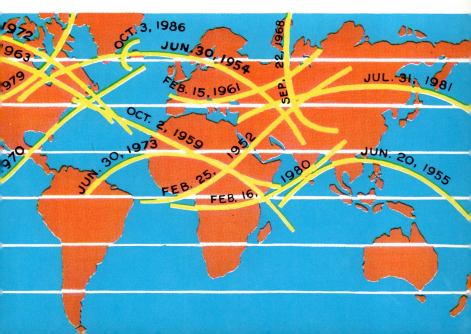

Total Eclipse of the Moon

LUNAR ECLIPSE • The earth's shadow is some 900,000 miles long. When the moon enters into it and is eclipsed, the eclipse lasts as long as several hours and may be total for as much as 1 hour and 40 minutes. In any year there may be no eclipses of the moon or as many as two and rarely three. Though there are fewer eclipses of the moon than of the sun, they last longer and can be seen by more people over a wider area. Because some of the sunlight striking the earth is diffused and scattered by our atmosphere, the earth's shadow is not completely dark. Enough of this light reaches the moon to give it a faint coppery glow even when it is totally eclipsed. An eclipse of the moon occurs only at the time of full moon. Because of angles of the moon's orbit, it may miss the

TOTAL LUNAR ECLIPSES

Date	Time of Midpoint of Eclipse (EST)	Duration of Eclipse	Duration of Totality
Jan. 30, 1972	5:55 a.m.	3 h. 25 m.	0 h. 40 m.
May 25, 1975	12:45 a.m.	3 h. 40 m.	1 h. 30 m.
Nov. 18, 1975	5:25 p.m.	3 h. 25 m.	0 h. 45 m.
Sept. 6, 1979	5:55 a.m.	3 h. 25 m.	0 h. 50 m.
July 6, 1982	2:30 a.m.	3 h. 45 m.	1 h. 40 m.
Dec. 30, 1982	6:25 a.m.	3 h. 30 m.	1 h. 5 m.
April 24, 1986	7:45 a.m.	3 h. 30 m.	1 h. 10 m.

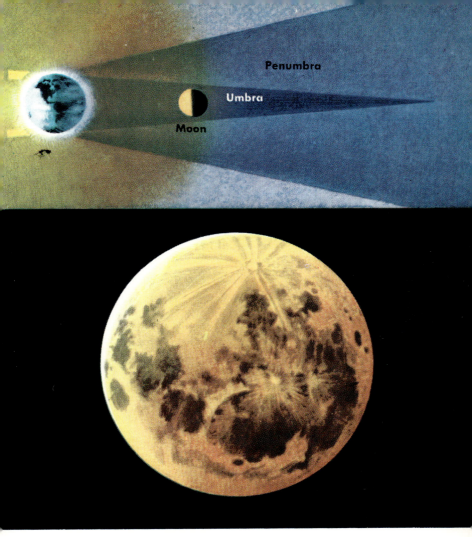

shadow of the earth completely at that time, or it may only pass through the penumbra. A lunar eclipse offers proof of the earth's shape, for the umbra that passes over the moon has the distinct curve of a shadow of a ball.

The table shows total lunar eclipses which can be seen in most parts of this country and southern Canada.

THE CONSTELLATIONS

The 88 generally recognized constellations are listed below. All but a few appear on the star charts in this guide. Keep in mind that latitude, season, elevation, atmospheric conditions, and time of night all determine whether a constellation is visible.

If a constellation is too far south to be seen from any part of the northern hemisphere, a figure is given which is the northernmost latitude at which the constellation (or most of it) can be seen under good conditions. Thus the figure "25°N." following Crux means that this constellation can be seen no farther north than about 25°N. latitude—the latitude of southern Florida.

The best season for seeing the constellation is indicated by the date on which the constellation reaches its highest point above the horizon (the meridian) at 9 p.m. Each evening a constellation reaches the meridian about 4 minutes earlier than the previous evening; each month, about 2 hours earlier. Thus Andromeda is on the meridian at 9 p.m. on Nov. 10; at about 8:56 p.m. on Nov. 11; at about 5 p.m. on Jan. 10.

Constellation	On Meridian 9 p.m.
Andromeda	Nov. 10
Antlia (The Pump) 50°N.	Apr. 5
Apus (The Bird of Paradise) 5°N.	June 30
Aquarius (The Water-bearer) 75°N.	Oct. 10
Aquila (The Eagle)	Aug. 30
Ara (The Altar) 30°N.	July 20
Aries (The Ram)	Dec. 10
Auriga (The Charioteer)	Jan. 30
Boötes (The Herdsman)	June 15
Caelum (The Burin) 45°N.	Jan. 15
Camelopardalis (The Giraffe)	Feb. 1
Cancer (The Crab)	Mar. 15
Canes Venatici (The Hunting Dogs)	May 20
Canis Major (The Great Dog) 65°N.	Feb. 15
Canis Minor (The Little Dog)	Mar. 1
Capricornus (The Goat, or the Sea Goat) 65°N.	Sept. 20
Carina (The Keel) of the ship Argo, which is no longer a constellation. 25°N.	Mar. 15
Cassiopeia (The Queen)	Nov. 20
Centaurus (The Centaur) 35°N.	May 20
Cepheus (The King)	Oct. 15
Cetus (The Whale)	Nov. 30
Chamaeleon 5°N.	Apr. 15
Circinus (The Compasses) 25°N.	June 15
Columba (The Dove) 50°N.	Jan. 30
Coma Berenices (Berenice's Hair)	May 15
Corona Austrina (Southern Crown) 45°N.	Aug. 15

Constellation	On Meridian 9 p.m.	Constellation	On Meridian 9 p.m.
Corona Borealis (Northern Crown)	June 30	Pavo (The Peacock) 20°N.	Aug. 25
Corvus (The Crow) 65°N.	May 10	Pegasus (The Flying Horse)	Oct. 20
Crater (The Cup) 70°N.	Apr. 25	Perseus	Dec. 25
Crux (The Cross) 25°N.	May 10	Phoenix (The Phoenix) 40°N.	Nov. 20
Cygnus (The Swan)	Sept. 10	Pictor (The Easel) 30°N.	Jan. 20
Delphinus (The Dolphin)	Sept. 15	Pisces (The Fishes)	Nov. 10
Dorado (The Goldfish) 25°N.	Jan. 20	Pisces Austrinus (The Southern Fish) 55°N.	Oct. 10
Draco (The Dragon)	July 20	Puppis (The Stern of the ship Argo)	Feb. 25
Equuleus (The Colt)	Sept. 20	Pyxis (The Compass) 55°N.	Mar. 15
Eridanus (The River) 70°N.	Jan. 5	Reticulum (The Net) 25°N.	Dec. 30
Fornax (The Furnace) 55°N.	Dec. 15	Sagitta (The Arrow)	Aug. 30
Gemini (The Twins)	Feb. 20	Sagittarius (The Archer) 60°N.	Aug. 20
Grus (The Crane) 35°N.	Oct. 10	Scorpius (The Scorpion) 55°N.	July 20
Hercules	July 25	Sculptor 55°N.	Nov. 10
Horologium (The Clock) 25°N.	Dec. 25	Scutum (The Shield) 75°N.	Aug. 15
Hydra (The Sea Serpent) 70°N.	Apr. 20	Serpens (The Serpent) 85°N.	
Hydrus 15°N.	Dec. 10	Caput (Head)	June 30
Indus (The Indian) 35°N.	Sept. 25	Cauda (Tail)	Aug. 5
Lacerta (The Lizard)	Oct. 10	Sextans (The Sextant) 85°N.	Apr. 5
Leo (The Lion)	Apr. 10	Taurus (The Bull)	Jan. 15
Leo Minor (The Little Lion)	Apr. 10	Telescopium (The Telescope) 35°N.	Aug. 25
Lepus (The Hare) 65°N.	Jan. 25	Triangulum (The Triangle)	Dec. 5
Libra (The Scales) 70°N.	June 20	Triangulum Australe (The Southern Triangle) 20°N.	July 5
Lupus (The Wolf) 45°N.	June 20	Tucana (The Toucan) 20°N.	Nov. 5
Lynx (The Lynx)	Mar. 5		
Lyra (The Lyre)	Aug. 15	Ursa Major (The Great Bear)	Apr. 20
Mensa (The Table Mountain) 5°N.	Jan. 30	Ursa Minor (The Little Bear)	June 25
Microscopium (The Microscope) 50°N.	Sept. 20	Vela (The Sails of the ship Argo) 35°N.	Mar. 25
Monoceros (The Unicorn) 85°N.	Feb. 20	Virgo (The Virgin) 80°N.	May 25
Musca (The Fly) 15°N.	May 10	Volans (The Flying Fish) 15°N.	Mar. 1
Norma (The Level) 35°N.	July 5	Vulpecula (The Fox)	Sept. 10
Octans (The Octant) 5°N.	Sept. 20		
Ophiuchus (The Serpent Bearer) 85°N.	July 25		
Orion 85°N.	Jan. 25		

OBJECTS FOR OBSERVATION

VISIBLE THROUGH THE YEAR (Middle North Latitudes)

Constellations: See seasonal charts, pp. 54-99. Learn first those with brighter stars.

Bright Stars: See list, p. 35, and seasonal charts. Identify star types by color (p. 37). Estimate magnitudes by comparison.

The Moon: Detailed suggestions on p. 148. Map, pp. 146-147.

Milky Way: Binoculars resolve much of milkiness into thousands of stars. Note dark nebulae between Cygnus and Scorpius, and star fields in Cygnus and Sagittarius.

Asteroids: Positions given in astronomical periodicals. Large asteroids, when near earth, can be spotted with binoculars.

Planets: See pp. 124-125. Use binoculars for 4 of Jupiter's moons and Venus' crescent; telescope for Saturn's rings. Positions of Uranus and Neptune given in astronomical periodicals.

Meteors: Table of showers, p. 130. Suggestions, pp. 129-133.

Mizar (star): In Big Dipper, p. 64. Note companion, Alcor.

VISIBLE DURING PART OF THE YEAR (Middle North Latitudes)

The celestial objects listed below are at the meridian at 9 p.m. standard time during the months indicated. They are visible in middle latitudes for one or more months before and after the months indicated. Their positions when observed depend on your latitude and the hour of observation.

January

Pleiades (open cluster): Near Perseus, pp. 82 and 86.

Hyades (open cluster): In Taurus, pp. 90-91, 93. Easy for naked eye.

M42 (Great Nebula): In Orion, pp. 90, 92. Impressive in binoculars.

Betelgeuse (variable red giant): In Orion, pp. 90, 92. Compare with Rigel (mag. 0.3, p. 92) and Procyon (mag. 0.5, p. 95).

February

M35 (open cluster): In Gemini, near Castor's left foot, pp. 65, 66.

March

M44 (open cluster): In Cancer, in center of "square," p. 64. Called Praesepe or Beehive. Splendid in binoculars.

May

Coma Berenices (open cluster): Between Leo and Boötes, pp. 64-65. Use binoculars.

July

M13 (globular cluster): In Hercules, pp. 72, 74. Faint, fuzzy spot. Binoculars show glowing cloud; telescope, individual stars.

M6 and M7 (open clusters): In Scorpius (pp. 72, 76), 5° northeast of tip of tail. Fine in binoculars.

August

Epsilon Lyrae (double star): In Lyra, 2° northeast of Vega, pp. 72, 75. Close pair. Telescope shows each is a double.

September

Albireo (double star): In Cygnus, p. 78. Small telescope reveals pair, orange and blue. Superb.

November

M31 (Great Spiral Nebula in Andromeda): See p. 85. Small, faint spot to unaided eye; glowing cloud in binoculars.

December

Double cluster in Perseus: In center of Perseus, p. 82. Faint, hazy patch becomes, in binoculars, a splendid spray of stars.

Algol (eclipsing variable): In Perseus, p. 86. Observe magnitude changes. Compare with Polaris (mag. 2.1).

INDEX

Bold type denotes pages containing more extensive information.

Achernar, 35
Albireo, 78
Aldebaran, 35, 37, **93**
Algol, 38, 39, 86
Alpha Draconis, 53, 58
Alphard, 63
Alpheratz, 84, 85
Altair, 35, 71, **79**
Amateur activities, 7-10
Andromeda (an-DROM-e-duh), 80-82, **85**
Antares (an-TAIR-eez), 32, 33, 35, 37, 70, **76**
Aquarius, the Water Carrier, 82, 100
Aquila (uh-KWIL-uh), the Eagle, 72, **79**
Archer, 72, **77**, 100
Arcturus, 35, 37, 61, **68**
Aries (AIR-eez), the Ram, 81, 100
Arrow, 71, 79
Asteroids, 102, **123**
Astronomy, 4, 50, 101
Atmosphere, earth's, 20-21, 24-27, **110**
Auriga (oh-RY-guh), the Charioteer, 87, 90
Aurora, 23, **24-25**

Betelgeuse (BET-el-gerz), 34, 35, 37, 71, 89, 92
Big and Little Dogs, 90, 95
Big Dipper, 51, 54, **56**
Boötes (boh-OH-teez), the Herdsman, 64, **68**
Bull, 33, 40-41, 88, **93**, 94, 100

Cancer, the Crab, **62**, 65, 100; cluster, 40
Canis (KAY-nis) Major and Minor, 90, **95**
Canopus, 35, 37, 99
Capella, 35, 37, 38, **87**
Capricornus, 83, 100
Cassiopeia (kas-i-oh-PEE-uh), 55, **59**, 61, 80
Castor, 34, 38, 61, **66**
Centaur (SEN-tawr), 97, 98
Cepheid (SEE-fid), 34, 39, 57, 60
Cepheus (SEE-fuhs), the King, 55, 59, **60**, 80-81
Ceres, 15, **123**

Cetus (SEE-tuhs), the Whale, 39, 81, 82
Charioteer, **87**, 90
Clusters, **40-41**, 62-63, 66, 74, 76, 77, 87, 94
Coal sacks, 45
Colors, 18-19, 26-27
Columba, the Dove, 89, 96
Coma Berenices (ber-ee-NY-seez), 40, 63, 64, 73
Comets, 102, **126-128**
Constellations, **50-101**; autumn, 80-87; key to, 61; list, 156-157; north circumpolar, 52-61; south circumpolar, 97-99; spring, 62-69; summer, 70-79; winter, 88-96; zodiac, 100-101
Corona Borealis (boh-ree-AL-is), Northern Crown, 70, 72-73
Corvus (KOHR-vuhs), the Crow, 63, 73
Cosmic dust, 42
Counterglow, 135
Crab, **62**, 65, 90, 100
Crater, the Cup, 64
Crow, 63, 73
Crown, 70, 72-73
Crux (CRUHKS), 45, 99
Cup, 64
Cygnus (SIG-nuhs), the Swan, 45, 46, 71, 72, 74, **78**

Day and night, 112-113
Declination, 71
Degrees, measuring by, 71
Delphinus (del-FY-nuhs), the Dolphin, 71, 72, 79
Demon Star, 86
Deneb, 35, 46, 71, 78
Denebola, 67, 69, 71
Dolphin, 71, 72, 79
Dove, 89, 96
Draco (DRAY-koh), the Dragon, 55, 58

Eagle, 72, **79**
Earth, 53, 104, **110-115**
Earthshine, 142
Eclipses, 149-155
Ecliptic, 100

Fishes, 81, 82, 100
Fomalhaut (FOH-mal-o), 35, **81**
Fraunhofer lines, 18-19

Galaxy, 42-45
Galileo, 28
Gegenschein, 135
Gemini (JEM-i-nee), the Twins, 61, 65, **66**, 100
Goat, 83, 87, 100
Great Bear, 40, 46, 51, 52, 54, **56**

Hare, 89, 96
Hercules, 33, 70, **74**; cluster, **40-41**, 74
Herdsman, 61, 68
Horse, Winged, 80-81, 84
Horse-head Nebula, 45
Hunter, 92
Hyades (HY-uh-deez), 93-94
Hydra, the Sea Monster, 63, 73, 90, 98

Job's Coffin, 71, 79
Jupiter, 10, 35, 102, 105, **118-119**, 125

King, 55, 59, **60**, 80-81
Kuiper, G. P., 106

Laplace hypothesis, 106
Leo, the Lion, 61, 62, **67**, 101
Lepus (LEE-pus), the Hare, 89, 96
Libra (LY-bruh), the Scales, **70**, 73, 100
Light year, 31
Lion, 61, 62, **67**, 101
Little Bear, 54, **57**
Little Dipper, 54, **57**
Little Dog, 95
Lyra (LY-ruh), Lyre, 72, **75**

Magellanic Clouds, 43, 97, 98
Magnitudes, 34-35, 51
Mars, 32, 35, 102, 104-105, **116-117**, 124
Mercury, 35, 102, 104-105, 108
Messier, Charles, 41
Meteors and meteorites, 10, 67, 102, 126-127, 129-133; observing, 10, 132; showers, 130

159

Midnight sun, 114-115
Milk Dipper, 71, 77
Milky Way, 40-45, 77-78
Mira (MY-ruh), 39, **81**
Mizar (MY-zahr), 34, **38,** 56
Moon, 13, 102, **136-148**; eclipse, 154-155; map, 146-147; see also planets

Nebulae (NEB-u-lee), 41, 42, 43, **45-47**, 58, 75 77, 81, 85, 92
Nebular hypothesis, 106
Neptune, 104-105, **122**
Northern Cross, 45, 71, 72, 74, **78,** 83
Northern Crown, 64, 70, 73
Northern Lights, 23-25
North Star, 34, 39, 50, 53, **56-57,** 58, 60, 61
Novae, 39, 46, 59

Observatories, 8, 28, 30
Observing, 5-7, 88, 132, 148, 158
Ophiuchus (ahf-i-U-kuhs), the Serpent-bearer, 70, 72
Orion (oh-RY-ahn), the Hunter, 77, 88, **92**; nebula, 45, 47, 92

Pegasus (PEG-uh-suhs), the Winged Horse, 80-81, 84
Perseus (PER-sus), 38, 40, 59, 80-81, **86**
Photography 10, 29, 89
Pisces (PIS-eez), the Fishes, 81, 82, 91, 100
Planetariums, 8
Planets, 12-13, **102-122,** 123; brightness, 34-35; table, 104-105; life on, 103; locating, 124
Pleiades (PLEE-yuh-deez), 41, 88, **94**
Pluto, 102, 105, **122**
Polaris (po-LA-ris), 34, 50, 53, **56-57,** 58, 60, 61
Pollux, 35, 61, **66**
Precession, **53,** 60
Procyon (PRO-see-yun), 35, 37, **95**
Pyramids, 4

Queen, 55, **59,** 61, 80

Radio astronomy, 29, **47**
Rainbow, 20-21
Ram, 81, 100
Regulus (REG-yoo-luhs), 35, 61, **67**
Rigel, (RY-jel), 35, 37, 71, 89, **92**
Right ascension, 71
Ring Nebula, 47
Rocket to moon, 138

Sagitta (suh-JIT-uh), the Arrow, 71, 79
Sagittarius (saj-i-TAIR-ee-uhs), the Archer, 42, 71, 72, **77,** 100
Satellites, 102; see also under planet names
Saturn, 35, 102, **120-121,** 125
Scales, 70, **73,** 100
Scorpius, the Scorpion, 45, 70, 76, 77, 101
Sea Monster, 63, 73, 98
Seasons, 114-115
Serpens, the Serpent, 70-71, 83
Serpent-bearer, 70, 72
Seven Sisters, 41, 88, **94**
Shooting stars—see Meteors
Sickle, the, 67
Sirius (SEE-ree-us), 31, 33, 35, 36, 89
Sky, color of, 26-27
Solar system, 14-15, **102-155;** components, 102-103; in galaxy, 42; movement, 33, 74; origin, 106-107
Southern Cross, 46, 53, 63, 97, 99
Spectrum, **18-21,** 36-37, 48-49
Spica (SPY-kuh), 35, 36, 38, 69
Stars, 31-49; brightest, 35; brightness, 32, **34-35,** 38-39, 48-49, 51; classification, 36-39; clusters, **40-41,** 62-63; 74, 87; colors, 33; density, 33, 76; distances, 31; double and triple, 35, 38-39; eclipsing, 38; energy, 32; giants

Stars (cont.):
and dwarfs, 14-15, 33, 48-49, 92; light, 32; magnitudes, 34-35, 51; motions, 33; names, 51, 89; numbers, 31; origin, 48-49; size, 32; spectra, 36-37; temperatures, 16, 22, **32, 36-37,** 45; variable, 10, 34, **39,** 60
Sun, **16-27**; brightness, 32; classification, 37; distance, 31; eclipses, 149-153; light, **18-19,** 26-27, 114-115; observing, 16; spots, 22-23, 25; tide, 114-145
Sundial, 5
Sunrise and sunset, 26
Swan, 71, 72, 74, **78**

Taurus (TAWR-us), the Bull, 33, 88, **93,** 94, 100; clusters, 40-41
Telescopes, 10, **28-30;** making, 9
Thuban—see Alpha Draconis
Tides, 144-145
Time, 112-115
Triangulum, the Triangle, 81, 85, 98
Trifid Nebula, 46
Twilight, 27
Twins, 61, 65, **66,** 100

Universe, the, 12-13
Uranus (U-ruh-nuhs), 34, 104-105, 122
Ursa (ER-suh) Major, the Big Bear, 40, 46, 51, 52, 54, 56
Ursa (ER-suh) Minor, the Little Bear, 54, **57**

Vega (VEE-guh), 35, 37, 53, 71, 75
Venus, 34, 35, 102, 104-105, **109,** 124
Virgo (VER-goh), the Virgin, 63, 64, **69,** 100

Water Carrier, 82, 100
Whale, 39, 81, **82**

Zodiac, 100-101
Zodiacal light, 134-135